In Kon-Tiki's Wake

In Kon-Tiki's Wake

The Archaeology of the Transoceanic Experiment

by

P.J. Capelotti

www.Fireshippress.com

In Kon-Tiki's Wake by P J Capelotti

ISBN-13: 978-1-61179-269-0 (Paperback)
ISBN -978-1-61179-270-6 (e-book)

BISAC Subject Headings:
TRA008000 Transportation / Navigation
HIS000000 HISTORY / General
SCI052000SCIENCE / Earth Sciences / Oceanography

Cover Photo by Carl Christian Raabe
Cover work by Christine Horner
Title page photo of Easter Island moai taken by and used with the permission of Braulio Rojas.
Illustrated by C.L. Devlin

Address all correspondence to:
Fireship Press, LLC
P.O. Box 68412
Tucson, AZ 85737
Or visit our website at:
www.FireshipPress.com

DEDICATION

For Jenny

If an object be set afloat in the ocean, as at the equator, it would, in the course of time, even though it should not be caught up by any of the known currents, find its way to the icy barriers about the poles, and again back among the tepid waters of the tropics. Such an object would illustrate the *drift of the sea,* and by its course would indicate the route which the surface-waters of the sea follow in their general channels of circulation to and fro between the equator and the poles.

—Matthew Fontaine Maury,
from chapter XVII,
"Tide Rips and Sea Drift,"
The Physical Geography of the Sea
(1855)

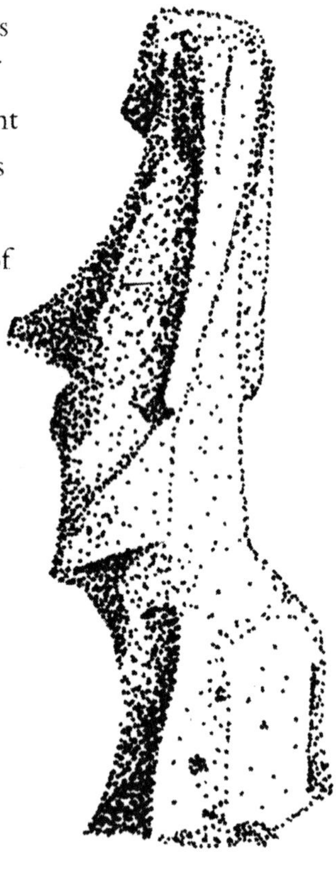

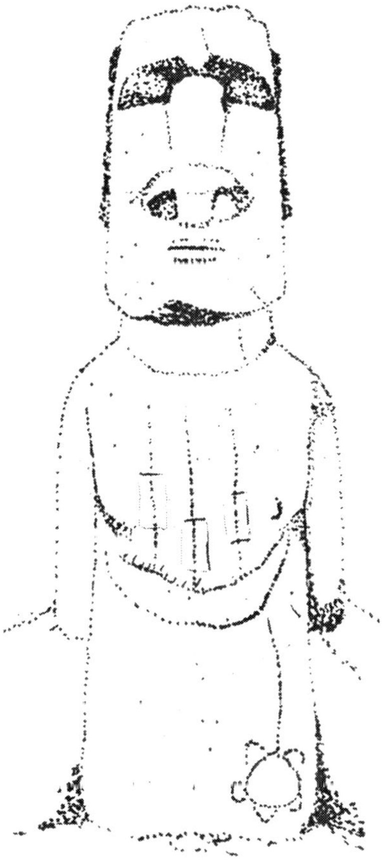

Books by P.J. Capelotti

Author

In Kon-Tiki's Wake: The Archaeology of the Transoceanic Experiment

Shipwreck at Cape Flora: The Expeditions of Benjamin Leigh Smith, England's Forgotten Arctic Explorer

The Human Archeology of Space: Lunar, Planetary, and Interstellar Relics of Exploration

God's Meadow: A Summer of Poems on the Edge of Oslofjord

Rogue Wave: The U.S. Coast Guard on and After 9/11

Sea Drift: Rafting Adventures in the Wake of Kon-Tiki

By Airship to the North Pole: An Archaeology of Human Exploration

The Wellman Polar Airship Expeditions at Virgohamna, Danskøya, Svalbard

Explorers Air Yacht: The Sikorsky S-38 Flying Boat

Editor

The Whaling expedition of the Ulysses, 1937-38

Life and Death on the Greenland Patrol, 1942

The Franz Josef Land Archipelago: E.B. Baldwin's Journal of the Wellman Expedition to Franz Josef Land, 1898-1899

The Svalbard Archipelago: American Military and Political Geographies of Spitsbergen and other Norwegian Polar Territories, 1941-1950

Our Man in the Crimea: Commander Hugo Koehler and the Russian Civil War

General Editor

U.S. Army Heraldic Crests

TABLE OF CONTENTS

List of Figures

PREFACE

When originally published in 2001 as *Sea Drift: Rafting Adventures in the Wake of Kon-Tiki*, this volume became the first book to collect the combined histories and anecdotes of every experimental drift voyage undertaken across the oceans of the world between 1947 and 1999. In 1999 alone, five different expeditions had sought to drift on their reed or balsa log rafts in search of the answers to perceived problems of prehistoric migration or navigation, making it the busiest year for transoceanic raft voyages since Thor Heyerdahl drifted across the Pacific on board his *Kon-Tiki* raft in 1947.

In the decade since, the number of experimental raft expeditions has dwindled. This is no doubt in part because the rise of the Internet has replaced much actual exploration with digital exploration. And Thor Heyerdahl, the inspirational figure behind so much of experimental voyaging, passed away in 2002 at the age of 87. Between 2000 and 2010, just one major balsa raft expedition, Torgeir Higraff's *Tangaroa* in 2006, sought to use the Pacific Ocean as an experimental laboratory. The story of *Tangaroa* was included in the 2007 Norwegian edition of *Sea Drift, Flåteferder i Kon-Tikis kjølvann* (Oslo:Flyt forlag).

Interest in such work rose again in 2012-13, with the release of the Norwegian biopic *Kon-Tiki*, a feature-length film that was nominated as one of the best foreign films of the year by the Academy of Motion Picture Arts and Sciences. So the time was right to produce a new edition of *Sea Drift*, this one revised and refocused for courses in archaeology. Where the original volume told the stories of these many expeditions chronologically, this new paperback edition from Fireship Press arranges them thematically, according to four key criteria for evaluating such archaeological experiments. It also assembles pedagogical materials at the end of the volume for use in adapting this material to courses in archaeological method and theory, as well as general introductions to problems in archaeology.

This volume also touches on some of the esoteric debates that did take place after *Kon-Tiki*, but that is not our main focus. The archaeology of the settlement of Polynesia is among the most vast and dimly understood areas of the whole field of anthropology. Entire libraries have been devoted to it, and no one person—save perhaps Hawai'i's legendary Kenneth Emory—could ever understand all of it. Perhaps because of its vastness, the Pacific invites large and sweeping hypotheses about the origins of its magnificent people. For those interested in the latest arguments regarding the archaeological study of the peopling of Polynesia, excellent sources already exist, prominent amongst these Patrick Vinton Kirch's *On the Road of the Winds* (California 2000).

In the decades following *Kon-Tiki*, the drifting expeditions that follow engaged in a kind of slow race across the seas and into the past. Many sought to emulate Heyerdahl, a few to challenge him. But whether praising or counter-punching, all used his methods: employing the prevailing ocean currents and winds as agents of cultural connections around the globe. Given that more than forty expeditions have followed these currents since *Kon-Tiki*, it is perhaps remarkable that only one was attended by loss of life, and that only when the expedition in question drifted more than a thousand miles beyond its original destination.

Yet even something as seemingly "eternal" as an ocean current or trade wind has become a prime focus of scholarly debate. As the great anthropologist Ben R. Finney has demonstrated through a series of experiments with Polynesian double-hulled voyaging canoes, "prevailing" does not mean "permanent." There are those, including Finney, who have either sailed successfully against the Pacific trade winds, or sat tight to await seasonal shifts in the prevailing winds in order to reach their destinations.

These expeditions, sailing rather than drifting, are not covered in this volume. They belong to the vast preserve of Pacific cultural revival triggered by Dr. Finney's applied research. Finney examines the efforts to rebuild the cultural database of Polynesian canoe sailing in several anthropological landmarks, notably *Hokule'a: The Way to Tahiti* (Dodd, Mead & Co., 1979), *Voyage of Rediscovery* (California 1995), and "Wait for the West Wind" (Finney, et al., 1989, J. Poly. Soc. 98).

Sitting over dinner with Ben Finney and his wife at an open-air restaurant along the Honolulu waterfront in the winter of 1999, I was struck by the vast and seemingly unbridgeable dissonance between sailors and drifters. The former tack against the wind, the latter must of necessity journey wherever the wind blows. In another sense, one fights nature, one submits to it. The sailor is heroic, the drifter, romantic. Even though the drifter often raises sail, he does so only with the wind directly—or very nearly so—at his back.

This is not in any way to suggest that one can drift on a balsa raft knowing nothing about the rudiments of sailing. It is merely to propose that we still know how to sail in a systematic way because sailboats are still in use all over the world. On the other hand, it is very likely that no one had attempted a transoceanic voyage on a raft for at least several hundred years prior to the *Kon-Tiki* expedition in 1947. Even after the more than forty raft expeditions described in this volume, we still cannot be certain how such craft were navigated in prehistory.

As Thor Heyerdahl himself wrote in *Kon-Tiki*, "the original Polynesian race must at some time, willingly or unwillingly, have come drifting or sailing to these remote islands." That the preponderance of evidence points now toward willing sailors rather than unwilling drifters is small indictment of those who did not have fifty years of sophisticated archaeological analysis upon which to base their theories.

The expeditions described herein took their inspiration and, in many cases, their technology from Heyerdahl's now-classic example. These expeditions did not race Heyerdahl in a literal sense, but they all compared the distances and speeds of their rafts to the rafts constructed by the Norwegian. Moreover, these expeditions were races against time, against dwindling supplies of food and water, against the inevitable deterioration of the balsa, bamboo, and reed rafts themselves. Most took place in the Pacific Ocean—though I describe or list all transoceanic raft expeditions on all the world's oceans—and one look at the bar graphs in the Appendices will show that the vast majority of raft expeditions used the Pacific to drift to glory or ignominy.

P.J. Capelotti
Abington, Pennsylvania
October 2000 and
October 2013

INTRODUCTION:

The Theory of the Raft in Experimental Archaeology

> What lies beyond the margin of the world often sings to us with the voice of a siren, as if calling us into its embrace. We listen, we are lured, and finally we are seduced. The heavily scored margins on charts that I have observed over the years are testament to this predilection on the part of many seafarers. They are utterly bewitched by the prospect of continuing along one rhumb line until it reaches its farthest point. They want to find out whether its ultimate destination concurs with their idea of how the world really is.
>
> —James Cowan
> *A Mapmaker's Dream: The Meditations of Fra Mauro, Cartographer to the Court of Venice*

On August 7, 1947, the most famous raft of them all crashed onto the windward reef of the Polynesian island of Raroia. Named *Kon-Tiki* after a legendary prehistoric sun god, the raft had drifted more than four thousand miles in one hundred and one days on its voyage from the old Peruvian naval port of Callao. Thor Heyerdahl, the Norwegian explorer who, with five companions, built and tested the raft, sought ultimately to understand how humans first inhabited the islands of the Pacific. The expedition was therefore both history's greatest sea adventure and archaeology's most original and famous experiment.

The arcing trajectory of the raft as it drifted from Peru was nothing less than a comet streaking across the previously calm and cloudless skies of Pacific archaeology. When the raft landed in Polynesia, it was as if that comet had smashed to earth, sending up clouds of debate. In the years to come, serious people would say that Thor Heyerdahl and the *Kon-Tiki* expedition did more to bring Pacific studies into the modern world than any other single event in history. For at the moment *Kon-Tiki* reached Polynesia successfully, all of the world's oceans, especially the Pacific, became archaeological seas, where people could explore not just the present and the future, but the remote past of our distant ancestors.

In the aftermath of Thor Heyerdahl's *Kon-Tiki* expedition in the summer of 1947, thirty-seven similar expeditions sought to employ Heyerdahl's method of recreated maritime technology in order to reach widely varying goals, with varying levels of relevance to archaeology and cultural diffusion. The motivations behind these goals can be ascribed to four distinct categories: science, pseudo-science, adventure, and survival. The twenty expeditions identified as having a core of scientific intent were those that most closely identified their method with the examples of *Kon-Tiki* (if a balsa raft expedition) and *Ra II* (if a reed ship expedition).

Of the two survival expeditions, one used a rubber boat and the other was created by salvaging water barrels carried on a previous raft. Of the nine expeditions whose motivation was exclusively adventure, five used balsa rafts and all operated in the Pacific. Of the six expeditions whose motivation can be demmed as pseudo-scientific, one used Chilean cedar, while four employed modern plywood and one even more modern fiberglass.

Motivations for experimental archaeology raft expeditions, 1947-2006

Science	Pseudo-science	Adventure	Survival
Kon-Tiki	*Lehi I*	*Seven Little Sisters*	*L'Heritique*
Tahiti-Nui	*Lehi II*	*La Cantuta I*	*Tahiti-Nui III*
Feathered Serpent	*Lehi III*	*L'Egare I*	
Ra	*Lehi IV*	*L'Egare II*	
Ra II	*Tahiti-Nui II*	*La Cantuta II*	
La Balsa	*Acali*	*Age Unlimited*	
La Tres Balsas		*Pacifica*	
Tigris		*Celeusta*	
Uru		*La Enduriencia*	
Hsu Fu			
Illa-Tiki			
Chimok			
Mata-Rangi			
La Manteña			
Abora			
Titi			
Kota-Mama			
Viracocha			
Tangaroa			
Total: 20	6	9	2

Of the total number of drift expeditions in re-created maritime technology that began with *Kon-Tiki*, eleven can be seen as producing significant results for studies of experimental archaeology and cultural diffusion. Of these eleven, only two—*Ra II* and *Tigris*—were not conducted in the Pacific Ocean.

Each of these eleven expeditions contributed to the evaluation of unique variables in prehistoric voyaging by watercraft essentially aligned with the prevailing winds and currents of the planet. The first three, *Kon-Tiki*, *Tahiti-Nui*, and *Ra II*, demonstrated the durability of long-distance watercraft constructed from three materials—balsa, bamboo, and [papyrus] reed—previously thought incapable of such range and endurance. The next three expeditions, *La Balsa, Tigris*, and *Hsu Fu*, more than doubled the range of the first three expeditions, and thereby demonstrated that these same materials—balsa, [berdi] reed, and bamboo—possess enough durability that they would have allowed prehistoric mariners to encircle the globe.

The next three expeditions, *Mata Rangi, Illa-Tiki*, and *La Manteña*, were expeditionary failures but experimental successes. The massive reed ship *Mata Rangi* broke apart soon after its launch from Rapa Nui, providing evidence to believe that reed ships of prehistory must have been built more within the parameters of *Tigris*. It also provided support for archaeologist Arne Skjølsvold's belief (personal communication, March 14, 2000) that the unique Rapa Nui moai statue known as No. 263 and used as the basis for *Mata Rangi* probably represents not a prehistoric Rapa Nui ship but a historic European sailing vessel. (If Skjølsvold is correct, then it is possible that the moai 263 carving shows not humans on the deck of a three-masted ship, but flames, and represents the burning of a European vessel. Depending on the amount of time it took for the downslope movement of soil to cover the lower part of the monolith, it is perhaps even a representation of the destruction of H.M.S. *Bounty* by Fletcher Christian and his fellow mutineers on nearby Pitcairn Island in 1789.)

The *Illa-Tiki* and *La Manteña* expeditions of John Haslett (Haslett 2006) demonstrated the very great struggle to successfully navigate a balsa raft northwards along the coasts of South America to Mexico. This route, speculated as an avenue for the transmission of the knowledge of metalworking in prehistory (Dewan and Hosler 2008), is fraught with dangers. These include oceanographic dangers such as the spiral of currents near Panama known as the Gyre, as well as the evident risk to balsa wood from the marine invertebrate *teredo navalis*.

The final two expeditions of experimental significance—the 2000 reed ship *Viracocha* and the 2006 balsa raft *Tangaroa*—extended our knowledge of the speed with which prehistoric mariners could have traveled across vast distances. *Viracocha* showed that a reed ship launched from the west coast of South America could reach Rapa Nui in as little as six weeks. This suggests the possibility of frequent round trips between the megalithic cultures of Rapa Nui and the reed ship building cultures of Lake Titicaca in Bolivia. Similarly, *Tangaroa* made the journey from Peru to Polynesia a full month faster than did *Kon-Tiki* in 1947.

Heyerdahl's voyage single-handedly opened the world's waterways to these archaeological experiments—both eminently rational and exceedingly bizarre—about prehistoric human expansion around the globe. The drifting raft became not only the essential component of a bewildering array of theories of possible human migrations or prehistoric contacts between widely scattered human groups, but of theories of individual human endurance.

Drifting rafts were used to show that the Chinese could reach British Columbia, that an old man can endure six months alone on the Pacific, and that humans could work out their sexual frustrations while cooped up on a raft in the middle of the Atlantic. One earnest captain decided that Mormons had colonized Hawai'i from Redondo Beach, California, and built a raft and drifted to Honolulu in 1958 to prove it. A French doctor settled himself in the bottom of an inflatable boat barely bigger than himself, and drifted from Europe to the Caribbean, all to prove that shipwreck castaways could survive by drinking minute quantities of salt water each day. Another Frenchman sailed in the opposite direction, with two human and two feline companions, to prove that a primitive log raft could sail west to east across the Atlantic, from Newfoundland to Ireland, in the amiable absence of evidence that such a voyage had ever been undertaken.

And so it went. Hypothesis after hypothesis piled upon raft after raft, all set adrift in hopes of fetching up on some sunlit shore where, their captains hoped, they would remarkably metamorphose into theories acceptable to largely disinterested academics.

The common thread in all these expeditions was their obligation to Heyerdahl's *Kon-Tiki*. *Kon-Tiki* transformed the world's oceans from the highways of merchants and navies, from the work places of fishermen and play places of sportsmen, into a new and almost chaotic kind of global scientific—and at times pseudo-scientific—laboratory. *Kon-Tiki* set this global sea drift in motion. And inside this new global laboratory, Heyerdahl himself, and many of those who followed him, sought nothing less than to understand the paths of human global exploration.

Chapter I

The Kon-Tiki Expedition

Thor Heyerdahl and the Seagoing Archaeological Experiment
Thor Heyerdahl and the Riddle of Easter Island
Critiques and Cautions: of Time and Seamounts
More Cautions: Follow the Bamboo

1. Thor Heyerdahl and the Archaeological Experiment

> Our intention was to test the performance and quality of the Inca raft, its seaworthiness and loading capacity, and to ascertain whether the elements would really propel it across the sea to Polynesia with its crew still on board.
>
> —Thor Heyerdahl
> *Kon-Tiki: Across the Pacific by Raft* (1947)

On a single narrow street in Oslo, Norway, called Bygdøynesveien, stand perhaps the most remarkable collection of maritime museums in the world. At the head of the way, isolated in space as its collection was in time, is the Viking Ship Museum. Here are housed two of the world's greatest maritime cultural treasures, the Oseberg and Gokstad Viking ships, both excavated in the late nineteenth century. If you leave the Viking Ship Museum and walk to end of Bygdøynesveien, past affluent homes roofed with gleaming glazed tile and sheltered behind high thick hedges, you emerge onto a rocky plain overlooking Oslo fjord.

A triangle of maritime museums surrounds a small turning circle. There is the great A-frame housing Fridtjof Nansen's *Fram*, the most famous polar research vessel in history. Just behind the Fram Museum, displayed out-of-doors, is the only slightly less well-known *Gjoa*, the almost tiny vessel in which Captain Roald Amundsen completed the first voyage through the Northwest Passage in 1906. A large building, shaped like the prow of a ship jutting into the fjord, houses the Norwegian Maritime Museum, with its collections covering Norway's naval, merchant, and whaling heritage.

Across the circle from the *Fram* and Maritime collections, as if set slightly apart from both, stands the "Institute for Pacific Archaeology and Cultural History," otherwise known as The Kon-Tiki Museum, the most popular museum in Scandinavia. On this street of institutions celebrating the Vikings, polar explorers, and whaling captains, a museum devoted to a raft expedition to Polynesia seems at first glance somewhat incongruous. And yet it is not.

When Fridtjof Nansen in the 1890s proposed drifting across the polar basin by setting the *Fram* into the same ice pack that had destroyed all previous expeditions in search of the North Pole, more than one observer saw disaster written all over the enterprise. Nansen's theory turned all previous experience on its head. When *Fram* and its crew emerged alive in 1896 after three years in the ice, with Nansen's theory vindicated, it was as if men long given up for lost had returned from a twilight world. It is easy to imagine that Thor Heyerdahl, challenged after the Second World War to demonstrate whether a raft made out of balsa wood could drift more than 4,000 miles from South America to the islands of eastern Polynesia, sought some comfort in the experience of his fellow Norwegian Nansen.

Heyerdahl's balsa raft, which did not sink to the bottom of the Pacific as so many had predicted it would, is now the centerpiece of perhaps the most unique museum in the world: the only institute anywhere that is devoted in the main to a single archaeological experiment. So it is not especially ironic that these two most famous of seagoing drifters occupy the end of the same street in Norway. Were they not safely enclosed within their permanent shelters, you could with a strong arm throw an obsidian mataa spear from the deck of Heyerdahl's *Kon-Tiki* and strike the double-hull of Nansen's *Fram*.

But *Kon-Tiki* was a drift expedition of a very different sort. It was conceived and executed as a scientific test of an anthropological hypothesis. Nothing of the sort had ever been done before. A seagoing archaeological experiment. It is difficult today, when almost no archaeology documentary appears on television without some sort of experiment into the behavior of a prehistoric culture, to recall just how unique this concept was in 1947.

It is true that the idea of experimentation within the field of archaeology had been explored as early as the 1840s. Scandinavian naturalists would observe how dogs crunched bird bones, in order to understand why the long bones of birds appear with such frequency on archaeological sites, while the rest of the skeleton does not. And other explorers had crossed various bodies of water in replica craft, seeking to imitate cultures of an earlier age—the 1893 Atlantic crossing by a replica Viking ship comes to mind.

But the Scandinavian archaeologists of the 1840s were seeking to understand the operation of a known, recorded phenomena. In a similar way, the Viking presence in Iceland and Greenland was a long-acknowledged fact. Few doubted the immense seaworthiness of the Viking longboat. Early journeys of the 200 miles that separate Greenland from Newfoundland had long been considered plausible and brief if arduous and untested. They were in fact spoken of within the Sagas. Then Helge Ingstad's and Anne Stine's work in the early 1960s demonstrated conclusive archaeological evidence for Viking encampments in North America and put the matter to rest.

Heyerdahl calibrated his ideas on far more enormous scales of time, endurance, and space. First, he sought to recreate a vessel he believed prehistoric peoples had invented and mastered nearly two millennia into the past. Second, he proposed to risk his life by making an actual voyage on board this vessel, which he believed could remain afloat for months. Finally, unlike Vikings closing the stormy 200-mile gap separating Greenland from Newfoundland, Heyerdahl proposed to use his recreated prehistoric raft to connect a continent with an archipelago separated by more than 4,000 miles of ocean.

Heyerdahl came to this idea after living as a Polynesian himself. Newly married, he spent a year in the Marquesas islands with his new wife in the late 1930s to research the zoological populating of the Pacific. Seeking to understand how animals reached remote islands, he recognized instead archaeological and oceanographic patterns that offered him the chance to turn his attention from animals to prehistoric peoples. As weeks and months passed into the complete cycle of a year, he became more and more transfixed by the infinite succession of waves breaking from the east, accompanied by a seemingly endless procession of clouds driven by southeast trade winds. The march of southeast wind and wave took hold of his imagination, and he imagined these same physical processes carrying cultural elements from the South American coast, 4,000 miles away, into the eastern rim of Polynesia.

His year in the Marquesas led Heyerdahl on a ten-year archival odyssey, to the chronicles of the European discoverers of the Pacific, to the myths of native Polynesians, and to such practical questions as how long a coconut can remain afloat and viable in salt water. He followed intently the discussions concerning the peopling of Polynesia. In 1941, Heyerdahl advanced the idea of a dual migration into Polynesia from the Americas. The first, he wrote, was led by a "pre-Incan civilization, with its centre near Lake Titicaca and along the Peruvian coast below, [which] seems to have swept the islands at a comparatively early period, via Easter Island," while a second wave arrived from the northwest coast of North America aided by northeast trades and currents.

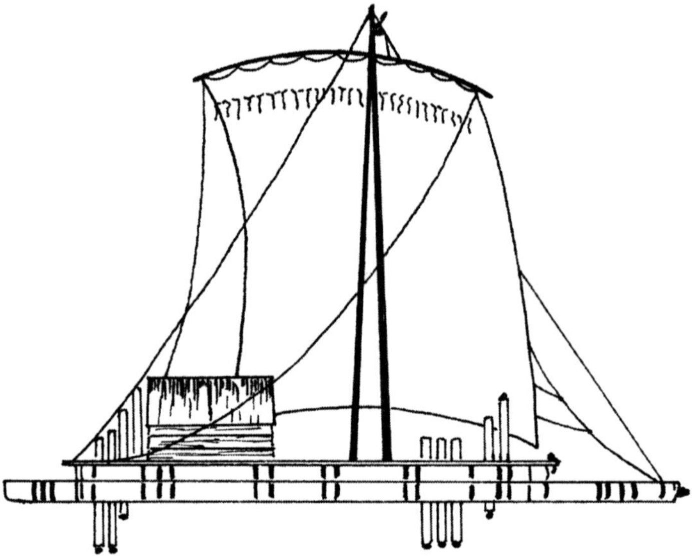

1. The sketch upon which Thor Heyerdahl, Vital Alsar, and John Haslett all based the construction of their balsa-wood rafts (after a sketch made in Guayaquil, Ecuador, by F. E. Paris in 1841–43)

Journeying to Ecuador with sketches of native rafts drawn by the earliest European explorers of South America, Heyerdahl harvested nine enormous balsa logs and floated them down the Palenque River to Guayaquil, and thence to the naval yard at Callao on the coast of Peru. The logs were lashed together to form the deck of the raft he would soon christen *Kon-Tiki*. With the raft complete, Heyerdahl and five crewmembers—two of them heroes of the Norwegian Resistance during the Second World War—departed Callao on 28 April 1947. One hundred and one days and 4,300 miles later, the raft smashed into the reef at Raroia in the Tuamotu Islands, and a debate joined that continues today.

2. Thor Heyerdahl and the Riddle of Easter Island

> We kept on sitting there and admiring the sea which, it seemed, was loath to give up demonstrating that here it came rolling in from eastward, eastward, eastward. It was the eternal east wind, the trade wind, which had disturbed the sea's surface, dug it up, and rolled it forward, up over the eastern horizon and over here to the islands. Here the unbroken advance of the sea was finally shattered against the cliffs and reefs, while the east wind simply rose above the coast and woods and mountains and continued westward unhindered, from island to island, toward the sunset.
>
> —Thor Heyerdahl
> *Kon-Tiki: Across the Pacific by Raft (1947)*

Thor Heyerdahl was hardly the first person to become obsessed with the riddle of Easter Island. Known as Rapa Nui, or Big Rapa, to its inhabitants, Easter Island has fascinated the western imagination since the moment Jakob Roggeveen's fleet sighted the place on Easter Sunday, 1722. Its famous moai statuary makes it one of anthropology's most enduring cultural enigmas. The easternmost inhabited island of Polynesia, it lies 2,400 nautical miles west of South America and some 1,400 miles east of Pitcairn

Island, its nearest inhabited Polynesian neighbor. In its extreme isolation, it has been populated solely by organisms able to travel thousands of miles by sea or air.

As part of his research into Polynesian migrations, Heyerdahl linked several cultivated plants from America with islands of Polynesia, including Easter Island. The most important was the South American sweet-potato (*Ipomœa batatas*) or, in the local name, kumara, a plant name, as one scholar noted, "that has stirred the imagination of scientists working in the Pacific like no other." The sweet potato is well established throughout Polynesia. As both a cultivated plant and a name, it spread throughout the Pacific islands by human contact.

That contact is conclusively aboriginal, the cultigen having been observed in New Zealand by Cook on his first voyage and on Easter Island by Roggeveen in 1722, and having been described by traditional history as being located in Hawai'i as early as 1250 C.E. Even proponents of European introduction found it unlikely that the plant could have spread so far in the few decades separating the voyages of Cook and Roggeveen with those of the 1560s of the Spaniards Mendaña and Quirós (Buck 1938, 313). In any case the kinds of long inter-island voyages necessary for the settlement of the expanse of Polynesia—and the introduction of kumara to the islands—had stopped sometime after the close of the fourteenth century (Hornell 1946, cited in Heyerdahl 1952, 431).

Heyerdahl invoked traditional myths of Easter Islanders to support his theory of a Peruvian origin for sweet potato, tobacco, and chili peppers on the island. The ethnologist Alfred Metraux quickly criticized him for using a native "myth to prove a thesis and [then using] the thesis to test the veracity of the same myth" (Metraux 1957, 227). Yet, however other cultivated plants may have arrived in Polynesia, the American origins of the sweet potato remain unchallenged. Current linguistic evidence points to the Cuna language spoken in northern Columbia as the origin of the word kumara, which in its various transliterations followed the sweet potato across the Pacific.

The combination of balsa rafts, sweet potatoes, and megalithic industries, all found along the coast of Peru and Chile prior to the first human movements into Polynesia, was intriguing. When a similar combination was found to have existed on Easter Island in the first centuries C.E., interest turned to a search for a method of cultural transmission. For Heyerdahl, the natural way to introduce kumara to the Marquesas, or Easter Island, was alive, in a pot, on board a balsa log raft.

Yet any notion that pre-Incan maritime cultures of South America could have influenced the area was dismissed out of hand. If a balsa raft could float for no more than a few days or weeks, all the historical or circumstantial evidence in the world suggesting otherwise was irrelevant. Scorned for six years, Heyerdahl realized that his ideas would never be taken seriously unless he could prove the seaworthiness of the pre-Incan balsa raft. So he turned to an area of inquiry not even named when he employed it: experimental archaeology, the use of recreated technologies to study the invention, use, and cultural implications of human ingenuity.

With the success of *Kon-Tiki*, Heyerdahl organized a pioneering expedition to Easter Island in 1955. With the archaeologists William Mulloy, Edwin Ferdon, Jr., Arne Skjølsvold, Carlyle Smith, and Gonzalo Figueroa, Heyerdahl was able to establish that the island was apparently already inhabited by C.E. 400. In his interpretation of the evidence, Heyerdahl saw a three-age sequence of occupation. An Early Period culture constructed solar observatories oriented toward equinoctial and summer solstice positions of the sun before a Middle Period culture of stone carvers placed the famous moai statues on ceremonial platforms, or ahus.

The Norwegian excavations seemingly demonstrated the existence of a fire pit/ditch, where the moai stonecutters, or "Long

Ears," were annihilated by a race of "Short Ears," late in the seventeenth century. Combined with his comparison of the moai statuary on Easter Island with the monoliths at Tiahuanaco near the shores of Lake Titicaca in South America, Heyerdahl was quick to see a link, and the link was made of balsa logs. As Heyerdahl often noted, the combination of balsa rafts, sweet potatoes, and megalithic industries, all occurring along the coast of Peru and Chile in a time prior to the first human movements into Polynesia made speculation about transoceanic contacts almost unavoidable.

But did Heyerdahl's drifters arrive on board simple balsa log rafts or sophisticated ships woven from freshwater reeds?

Mythological scenes from Peruvian prehistory before the highland Incas conquered the coastal region depict, not balsa log rafts, but reed boats. Heyerdahl compiled a massive case, published as American Indians in the Pacific, that both highland mariners on Lake Titicaca and pre-Incan Mochica mariners of coastal Peru navigated large and sophisticated ships constructed from totora reeds. These are the same American fresh-water reeds (*Scirpus riparious*) discovered in the three crater lakes of Easter Island. Natives who witnessed the petroglyphs and carvings of ships unearthed by Heyerdahl's excavations on Easter Island "were positive that the ships ... represented the great vessels that their ancestors had built of totora reeds."

Heyerdahl cannot be faulted for selecting the balsa log raft for his experiment in 1947. It was not until 1970, after one failed attempt, that he himself proved the transoceanic capabilities of the reed boat, by sailing from Africa to the Caribbean in a craft of identical construction to those on Lake Titicaca (Heyerdahl 1971). In 1978, he built an even larger reed ship. This vessel, the *Tigris*, managed to stay afloat for five months, on a voyage from Iraq to Pakistan and then Africa.

But the drift voyage across the Atlantic was exactly that, a drift. If the sweeping Humboldt Current does not enable one to drift directly to Easter Island, how did prehistoric reed boats make the voyage? Heyerdahl sought an answer from the historical record. In 1828, the navigator J. A. Moerenhout, sailing from Chile, made the first of two cruises that passed by Easter Island. Studying the currents around the island, Moerenhout was emphatic in his belief that any future voyagers to Easter Island should start from northern Chile and utilize the free push of the southern branch of the Humboldt Current, which drifts almost directly to the island (Heyerdahl 1989, 60-61).

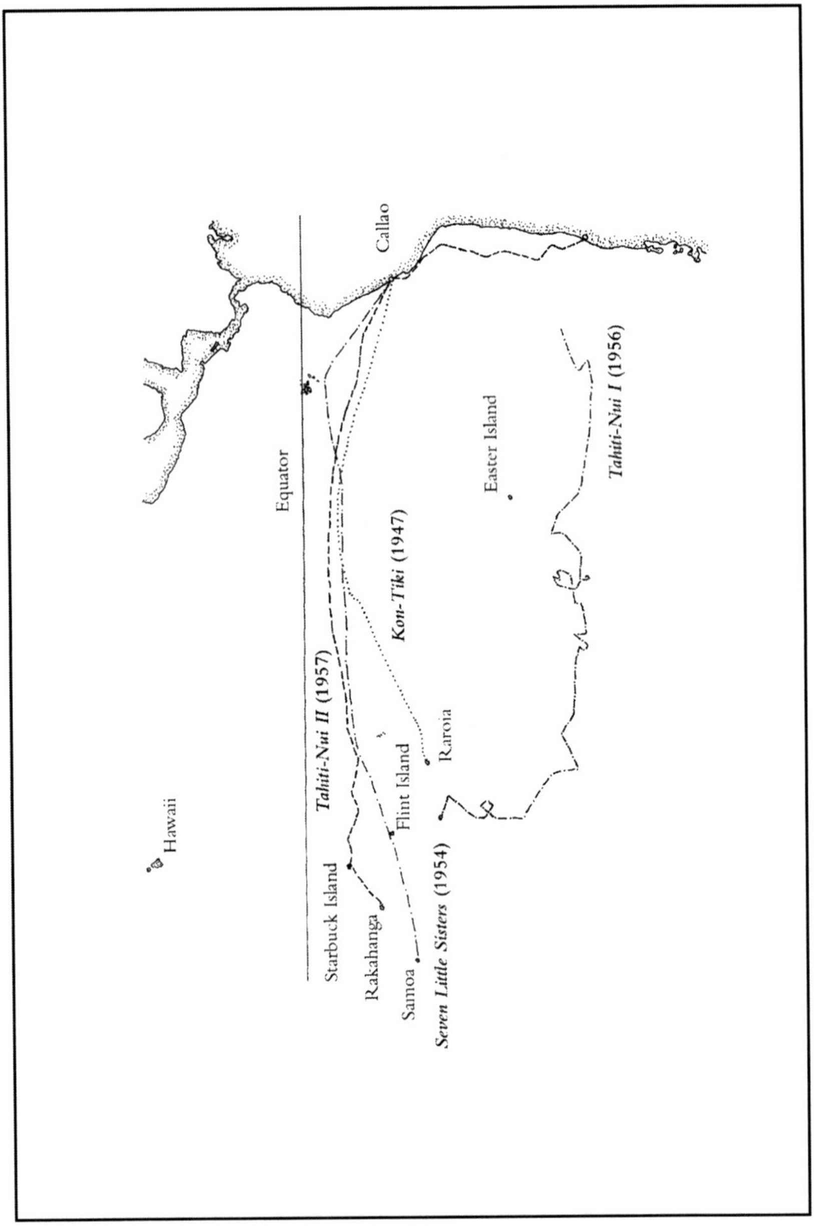

Fig. 2. Routes of the early transpacific raft voyages (after Heyerdahl 1979; Willis 1955; and Danielsson 1960)

This, of course, would presuppose a kind of Pacific super port in northern Chile, where prehistoric reed boat mariners would set their courses to the strange island 3,000 miles to the west. Such a port would have been far to the south of the Moche culture area but almost directly seaward of the mariners of Lake Titicaca, so the idea is not implausible. Like any hypothesis, it begs for experimental data, for aerial or satellite surveys of northern Chile in search of such a port. Heyerdahl investigated such reed boat ports in the Indian Ocean, and his colleague William Mulloy rebuilt the canoe or raft landing ramp at Tahai on Easter Island, which could have been used as a loading area for long-distance ocean voyages.

Such examples of sophisticated infrastructure seem to indicate well-established maritime cultures accustomed to regular sea trade. Mulloy's ramp at Tahai indicates almost certainly that the maritime people of Easter Island moved their giant statues or their topknot headdresses made of red volcanic scoria by sea, letting wind and tide do the work of human muscle. It is only modern man, with his limited imagination, that insists the Easter Islanders must have dragged these statues over miles of hills, only to haul them back to the sea and erect them on coastal temples. Fortunately, this is a fairly simple hypothesis to test, like our earlier proposed submarine explorations of the seamounts and guyots, or the search for a reed boat port in northern Chile. It is entirely conceivable that an underwater archaeology survey around the coast of Easter Island will reveal that one or more moai statues fell off the rafts transporting them around the island, and now rest at the bottom of the ocean.

Heyerdahl suggested another answer for the course of drift voyages from South America to Easter Island. "In normal years the main current by-passes Eastern Polynesia north of Easter Island and south of the Marquesas, and hits the islands in Polynesia further west... Normally it is only the outer sweep of the Humboldt Current that curves southwest and hits Easter Island, but in Niño years the full force of the Humboldt Current sweeps directly upon Easter Island. These are the years when the north coast of Peru is struck with catastrophic floods and thousands of coastal fishermen have to flee their homes with their families. During the last big Niño year in 1983 about 600 balsa rafts were seen far from their homes, paddling in the flooded desert around Tucumé" (personal communication). Was the great sun god Con-Ticci Viracocha swept with his people across the Pacific by a climatic twist of great ocean currents?

3. Critiques and Cautions: of Time and Seamounts

When his ship goes down a man's whole universe goes with it.
—Alain Bombard, 1953

The archaeological experiments of Thor Heyerdahl might be some of the most heavily criticized in the history of science. This criticism takes several forms. Some believed that the sail-equipped balsa raft had only been invented after the arrival of the Spanish in the 1500s; or that prehistoric peoples of coastal North America were shallow water sailors who preferred to hug the shore and keep land constantly within their sight. Some felt Heyerdahl would never survive on his raft for more than three months, three weeks even: his crew would die of thirst, or be swept overboard by storms, or be devoured by sharks. Worse, Heyerdahl proposed to close the gap between Peru and Polynesia by drifting with the Humboldt Current. This would take them out of all shipping lanes into some of the most remote waters on earth. If an emergency arose, they could count on no one to rescue them.

But by far the most common certainty was that the raft would sink within a few weeks of being launched. All knew, or thought they did, that pre-Incan balsa-wood rafts could not remain afloat long enough to reach even those islands of eastern Polynesia, such as the Marquesas, closest to the South American coast. After assembling mountains of anecdotal, botanical, and linguistic evidence bearing on his theory, Heyerdahl nevertheless found it rejected out of hand because few shared his belief in the watertightness of balsa logs. If a raft could not float for more than a few days or weeks, all the historical or circumstantial evidence in the world suggesting otherwise was rendered null.

In the end, his calculation worked brilliantly, and perhaps no one was more surprised than Heyerdahl himself. Like his contemporary, undersea explorer Jacques-Yves Cousteau, Heyerdahl suddenly found himself the very model of the modern, scientific, positive, rational romantic. Through their global explorations, both Cousteau and Heyerdahl achieved a kind of world-citizen, international icon status. And like Cousteau, Heyerdahl was dismissed, often unfairly and with extreme prejudice, by the elites within his own field of study. Oceanographers and marine biologists often sneered at Cousteau's popularization of undersea research even as they lined up for the grants such popularization engendered from the taxpaying public.

Heyerdahl's popular explorations of prehistoric migrations never led to the kinds of massive public funding for experimental archaeology that oceanography received following Cousteau's pioneering work in the 1950s and early 1960s. Royalties from books and movies did, on the other hand, provide the basis for Heyerdahl to carry out independent archaeological research and experiments across more than half a century.

More than twenty million copies of *Kon-Tiki* were sold in sixty-five languages. Like most classics, it seems almost ridiculously simple. Five young Norwegians and a Swede, adrift upon blue Pacific currents for an epic 101 days. The author, a young scientist, his ideas ignored by the academy, takes them instead across the high seas. His expedition drifts away from critics and coastline until, by bravery and wit and the loyalty of his mates, both idea and raft smash onto a coral reef thousands of miles away.

The intensely fascinating story and its almost breezy telling offered a world exhausted by war the enduring simplicity of a wooden raft drifting slowly across the mythical Pacific of Cook,

Bougainville, Banks, Roggeveen, and Fletcher Christian. Half a century later, the expedition and the incomparable book Heyerdahl wrote about it endure as pivotal individualistic triumphs of the modern era.

Yet for all the adventure it continues to represent, *Kon-Tiki* was only the first of many attempts by a young scientist of extraordinary courage to work out the implications of his ideas. Thor Heyerdahl began his work at a time when anthropologists, as his colleague Edwin Ferdon once wrote, were "essentially landlubbers," unwilling, so to speak, to get their feet wet. "Well, you can try a trip from Peru to the Pacific islands on a balsa-wood raft," were the icy words with which Herbert Spinden of the Brooklyn Museum in New York had dismissed Heyerdahl from his comfortable office. After *Kon-Tiki*, such scholarly passivity became untenable. For whether or not he ultimately demonstrated his vision of the seagoing capabilities of our exploring ancestors, Thor Heyerdahl most assuredly demonstrated that an anthropologist better be ready to take his hypotheses to sea.

Having demonstrated that Polynesia lay within the range of at least one type of aboriginal craft from South America, Heyerdahl found himself attacked on numerous other fronts. As an experiment, the *Kon-Tiki* expedition was criticized as a mere one-time success, lacking the controls and recording that would permit replication of the experiment as a controlled comparison. Yet, though Heyerdahl himself never recreated the entire experiment, others did. They are forgotten today, but between 1947 and 1973, more than a dozen manned rafts set out from Peru to cross the eastern Pacific to Polynesia, and all succeeded.

Major studies of native navigational abilities were conducted in part to show that native Polynesian navigation—as opposed to occasional pre-Incan rafting—was no accident of unlucky mariners. Gladwin, in his study of Puluwat navigation and logic (1970), found in the Caroline Islands that any notion of shipboard heuristic devices is absent. Once a system of navigation has been learned, it does not need to be modified to meet new exigencies because there are no new exigencies: the stars appear in the same place each night, and when the stars are hidden by cloud then the patterns of the waves will substitute for the stars. Innovation is not required, so it is not forthcoming.

Computer simulations of the peopling of Polynesia were produced. These provided mathematical models showing the relative probability of deliberate as well as drift voyages, primarily in western Polynesia. But they offered no hope to those who sought statistical backing for connections to the Polynesian outliers like Easter Island (Ward, Webb, Levison 1976, 66). In any case, if the settlement of Easter Island was a single random event buried in prehistoric time, any search for statistical pattern is bound to reveal nothing.

Across the whole of the Pacific, David Lewis (1972) concerned himself with retrieving, as far as possible, ancient and in all likelihood very closely held deepwater land-finding procedures. In part Lewis sought to alter the dim notions of primitive sailing and navigating abilities put forth by the New Zealander Andrew Sharp (1963), among others. The prehistoric deepwater capabilities of the Polynesian double-hulled voyaging canoe were subject to much doubt. Such doubts set anthropologist Ben Finney and his colleagues to construct replicas of such vessels in the 1960s and 1970s. Between 1976 and 1987 their major experiments in the double-hulled canoe, Hokule'a, saw this remarkable vessel travel over 20,000 miles of the Pacific: twice between Hawai'i and Tahiti and once on a round-trip between Hawai'i and New Zealand, or Aotearoa. Two legs of this last voyage, those from Ofu in American Samoa to Raratonga and from Raratonga to Tahiti—both concluded only with excruciating difficulty—were used by Finney (1993) to counter Heyerdahl's notion of "permanent trade winds" that act as an unalterable barrier to eastward canoe voyages.

But fighting these southeast winds and currents could not have been simple or easy for deep-water Polynesian ancients. And while difficult is not the same as impossible. Lewis, for one, makes the point that none of the relatively small sailing canoes he is familiar with sailed very close to the wind, and adds that there "seems no reason to suppose that the larger craft of yesterday were handled any differently" (Lewis 1972, 269). It is more likely, as Finney himself suggested, that voyagers from the western Pacific simply did not bother to fight prevailing winds. They rather stalled their expeditions to await the winter months, or the monsoons, both of which bring changes in the winds more favorable to west to east travel.

All of these considerations of wind and wave only intensify the further one ventures into the remote reaches of eastern Polynesia. And it was here, on Easter Island and islands further west, like Rapa and Raivavae, that Heyerdahl made his stand for South American contact with eastern Polynesia. Heyerdahl made much of the fact that *Kon-Tiki* drifted twice as far as would be necessary to reach Easter Island. Yet, until 1999, when Hokule'a accomplished it, none of the Pacific drifting or sailing experiments had come within a thousand miles of that elusive southeast corner of Polynesia. And Heyerdahl's hypothesis hinged on the relevance of raft voyages in explaining a solitary fact: the occurrence of the sweet potato on Easter Island.

There are no direct written records to document pre-Incan raft expeditions into the Pacific. That is, of course, the very nature of a prehistoric society. With no written documents, researchers are left to sort through the material creations people leave behind, everything from pyramids to the remains of last evening's meal. Linking prehistoric societies via maritime routes is perhaps the greatest challenge of all for an archaeologist. Prehistoric vessels, whether made of reeds or balsa wood or koa trees, have almost without exception vanished from the earth. Old vessels rot, and even a new vessel setting out from its homeport can be swept from the seas by storm, accident, or navigational incompetence.

To make matters worse for archaeologists trying to trace sea-going migrations, sea levels have risen dramatically, as much as 300 feet since the end of the Pleistocene era ten thousand years ago. Coastal cultures like the Paijan of the north coast of Peru, with their dramatic needle nose stone projectile points, may have done more than spear fish in tidal pools. It is possible that they could have ventured into the Pacific, a Pacific that because of vastly lower sea levels would have been geographically unique from the one we know today. But it is more than likely that the majority of Paijan camps or settlements, whether along coastal Peru or, conceivably, the fringes of some Polynesian island, were inundated millennia ago. Because of this, and with several notable exceptions, most archaeologists have been arguing over the wrong coastal geography for years.

Paijan sites on the north coast of Peru, like the site of Monte Verde in southern Chile, have been dated to between 7,000 and 12,500 years ago. Not until archaeologists investigate coastal shelf areas down to more than 300 feet below the surface, can we deny with certainty the possibility that human populations carried out voyages of exploration and migration thousands of years before we now believe it to have been possible.

Such "stepping stone" voyages would have taken advantage of vastly increased continental shorelines, as well as hundreds of now-submerged seamounts and guyots in the Pacific, which the lower ocean levels of the last ice age would have exposed and perhaps made habitable. Long chains of such volcanic seamounts are clustered across the middle of the Pacific at around 20° north, and additional clusters stretch northwest from San Francisco to Alaska, and southeast from Tahiti to Chile. A single cave on one of these submerged islands showing evidence of human occupation would rewrite the entire sequence of human global migrations.

This thought applies with equal force to the other side of the Pacific. Human ancestors reached Australia from Asia over water, because at no time in the last 160,000 years was Australia connected by land with Asia, even during the most severe ice ages. So there is no question that our ancestors were maritime explorers. At least one researcher believes that early human populations did not sail when the water was low, land was plentiful, and the next island or the clouds above it could be seen on the horizon. Rather, humans were forced to take to the sea in search of new land when the ice caps were melting, the oceans were rising, and available lands shrinking.

In the absence of submarine investigations of seamounts and guyots, archaeologists have been forced to invent a 25,000-year gap in the course of human movement across the Pacific. Early *Homo erectus* populations migrated into Southeast Asia hundreds of thousands of years ago, yet, prevailing thinking goes, were stopped by the deep waters of the Wallace Line from venturing to the combined landmass of Australia or New Guinea. With apparently greater seagoing technology, and despite rising sea levels, our Homo sapiens ancestors not only crossed the Wallace Line to Australia and New Guinea but then continued on to explore and colonize virtually every island from Guadalcanal to Easter Island.

Thor Heyerdahl himself speculated that early voyages from South America to Polynesia occurred around 500 years after Christ, and a thousand years before Spanish conquistadors extracted Incan accounts of great fleets sailing from the coasts of Peru to islands beyond the setting sun. So he picked up scattered clues in the archaeological record—a kind of stone wall here, a birdman cult there—and used the great ocean currents of the world to link them together in a kind of anthropological unified field theory of human maritime migration.

4. More Cautions: Follow the Bamboo

> And to think that it should come to this. That I should leave [the] poor old bones [of my ship] sticking on a reef as though I had been a damned fool of a southern-going man who must have half a mile of water under his keel to be safe! Well! Well! It's only those who do nothing that make no mistakes, I suppose. But it's hard. Hard.
>
> —Joseph Conrad
> *The Outcast of the Islands*

The imitative focus on the Heyerdahl-inspired balsa and reed ships has almost entirely obscured other, potentially more valuable vessel types and natural materials from the body of experimental nautical archaeology. The choice for future archaeological experiments in recreated technology is obvious: bamboo.

Only two transoceanic bamboo voyages have been undertaken and both with the object of reaching the Americas. The idea that Asian mariners could have crossed the Pacific two thousand years ago naturally leads one to consider the very origins of maritime travel. Could late Pleistocene-era bamboo rafts have sailed or drifted out from the low-lying settlements of continental Asia and across to the northern coast of New Guinea? Who was the first human to take to the sea? What did their vessel look like? Where were they going? Did they know—could they see—where they were going? How would one explore this idea experimentally? Localized voyages should be able to provide experimental data for the evaluation of innumerable and fundamental questions of Pleistocene human migrations.

During an archaeological expedition to the Indian Ocean archipelago of the Maldives Islands in the mid-1980s, Thor Heyerdahl himself spotted a large, abandoned bamboo raft in a lagoon at the island of Viringili. This particular type of raft was foreign to the Maldives, and locals told Heyerdahl that some fishermen had discovered the raft the day before, on the outer reef off the east coast of the island. Heyerdahl arrived at the bamboo vessel just as it was being torn apart in order to sell the pieces at a local auction. "It was a raft indeed: large, and built from giant bamboo. I had never seen bamboo as thick as this. Thick as telegraph poles and tied together in three layers under eight cross beams of wood with another layer of thick bamboo as a deck well above the water… Twelve meters long and three wide (39 feet by 10 feet approximately) with a freeboard of 40 centimeters (16 inches), the whole structure recalled the balsa raft *Kon-Tiki* which had been large enough to carry me and five companions across the Pacific" (Heyerdahl 1986: 188).

The only clues as to the origins of the raft were two paper labels, which Heyerdahl later identified as Burmese candy wrappers. The raft had most likely therefore floated nearly 2,000 miles from Burma, through the Bay of Bengal, and into the Indian Ocean to the Maldives. From his own rafting experience Heyerdahl looked at the barnacles growing on the bamboo and estimated that the vessel had been at sea about two months, pushed ahead of the Northeast Monsoon.

How such bamboo vessels could have carried the migrations of early peoples is a matter of intense interest to archaeologists. J.B. Birdsell identified five different routes from the ancient Southeast Asian continent of Sunda to the combined Australian-New Guinea continent of Sahul (Birdsell 1977). It seemed probable to Birdsell that "the watercraft used in the late Pleistocene were superior to those found in recent times in Australia and Tasmania. It is highly probable that there was a constant if somewhat straggling trickle of small groups of human beings over all or most of the routes. The size of the watercraft likely to have been used suggests that the groups consisted of small biological families" (ibid, 123).

Birdsell identified thirty-five different species of bamboo along these routes, indicating both the possible raw materials that prehistoric mariners may have used to construct rafts and the paths for which they were set adrift. Such varieties of both routes and raw materials provided scores of testable hypotheses. As Birdsell noted, assuming these many species of bamboo existed during ice age times of lower sea levels, they offer "a useful role for experimental archaeology" (ibid. 144). One could readily imagine a fleet of different bamboo rafts exploring dozens of possible routes for early humans in coastal Asia, especially in the lightly-studied Bay of Bengal, to Sahul.

There is already a model for such voyages. In January of 2000, a dozen men led by Robert G. Bednarik paddled a forty-foot bamboo raft called *Nale Tasih 4* from the east coast of the Indonesian island of Bali across eighteen miles of open ocean to the neighboring island of Lombok (Bednarik 2003; Bower 2003). This expedition continued Bednarik's experimental attempts to support a theory that *Homo erectus* and not *Homo sapiens* was the first hominid species to possess both the tools and maritime technology to cross to Wallace Line, some 800 kya (thousand years ago). In 1998, in a fifty-eight-foot-long bamboo raft, Bednarik and crew sailed from Timor to Australia, 600 miles in thirteen days. Bednarik's experimental nautical archaeology suggested that an optimal length for a bamboo raft was, like the bamboo raft found in the Maldives by Heyerdahl, about forty feet. This fascinating series of experiments connecting Indonesia to Australia should lead scholars of prehistoric population movements to a focus on underrepresented yet equally significant maritime domains such as the Bay of Bengal and the Andaman Sea.

CHAPTER 2

THE FIRST EXPERIMENT AFTER KON-TIKI: SURVIVAL

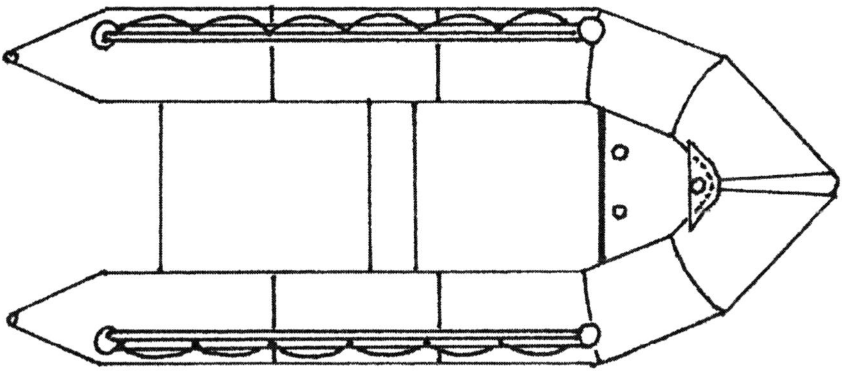

Fig. 3. Alain Bombard's inflatable transatlantic raft *L'Heretique* (after Bombard 1953)

L'Heritique (1952)

It was only a matter of months after the publication of *Kon-Tiki: Across the Pacific by Raft* in 1950 that the first raft to follow Heyerdahl's example set out across an ocean. But although this first ersatz *Kon-Tiki* used the same ocean current "conveyor belts" to reach its destination, in this case the raft, the theory, and the ocean used were all different from the balsa raft used to study prehistoric peoples of the Pacific. The goal was simple: survival.

A French doctor by the name of Alain Bombard was called one night in the spring of 1951 to assist the survivors of a trawler wreck. Instead of ministering to the stricken sailors, Dr. Bombard was instead shocked to discover that all of the crewmen had died. As he wrote later: "An error of navigation lasting a few moments had caused forty-three deaths and orphaned seventy-eight children (Bombard 1953, 9)."

To Bombard this disaster was a profound shock, as well as a challenge. It caused him to think deeply upon the nature of shipwrecks and their survivors, who often later became additional unfortunate victims through vicious combinations of starvation, thirst, and demoralization. By the fall of 1951, Bombard had arrived at the archives and laboratories of the Museum of Oceanography in Monaco, the same museum where Jacques-Yves Cousteau would later serve as director for thirty years. In Monaco, Bombard combined his lab experiments in the comparative compositions of seawater and the water he managed to literally squeeze from various species of fish, with historical research into the nature of previous shipwrecks and the lessons he might learn from them.

The results of his historical research led Bombard to conclude that: "Thirst kills more quickly than hunger, but despair is a greater danger than thirst (ibid, 30)." As a doctor, despair was a difficult thing for Bombard to quantify, even though he learned that ninety percent of shipwreck survivors died within three days, even though a human being could survive far longer than that even if completely deprived of food and water. If one could overcome despair, Bombard reasoned—and he was under few illusions as to this difficulty—then the only remaining problem was slaking thirst.

But thirst was another matter. Studying the components of various fish a castaway might encounter if shipwrecked somewhere in the Atlantic Ocean, Bombard learned that most of them were composed of eighty percent or more of water. Using a fruit press, Bombard learned how to extract this 'fish water,' and then disciplined himself to the point where he could drink it. He frankly admitted that this was hardly an easy task for a Frenchman. "Have you ever had to eat fish which some careless cook has forgotten to season? It is completely insipid (ibid, 24)." Bombard discovered that a shipwreck survivor would need to squeeze this water from the flesh of six or seven pounds of fish each day in order to provide enough fresh drinking water for survival.

But what if the fish were not biting? To solve this problem, Bombard reversed hundreds of years of common teaching, when he proposed that a castaway could in fact drink limited quantities of seawater and not damage one's kidneys. Comparing the composition of seawater with that of the bottled water from several French springs, he found that, other than the salt content, the other minerals in seawater existed in quantities no greater and in many cases in lesser amounts than bottled water on store shelves. If one used the salt in seawater as a substitute for an average daily amount of table salt, Bombard proposed that shipwreck survivors would not overwhelm the mineral salt filters in their kidneys.

Another transoceanic rafter who set out to cross an ocean a year after Bombard claimed that sailors of old had always known they could drink seawater, and not just when in extremis. As he wrote when he was halfway across the Pacific:

"For a number of days I had felt a bit stuffy around the middle and sluggish mentally, and I drank some seawater—about half a mugful, approximately the equivalent of an average cup. Since I always believed in the medicinal as well as nutritive value, it went down easily enough, and I made up my mind to drink it every day from then on. My technique of drinking was to pour it quickly down my throat and then inhale quickly a few times with open mouth till the gall and salt taste had disappeared (Willis 1966, 80)."

So fish water, supplemented with a pint and a half of seawater, could solve the problem of a shipwreck survivor dying of thirst. Bombard calculated that these innovations alone could help save many of the estimated 50,000 people who died every year clinging to lifeboats after a shipwreck. Bombard now had a reasonable hypothesis, but without a practical test it meant little. It would take some kind of real-life experience to demonstrate the value of his findings to skeptical fishermen and sailors. As he wrote, in a precise scientific way: "If my theory was to be something more than a hypothesis, if it was to serve some real purpose, it was essential to reduce the experiment to human terms in an actual sea voyage. I had to find some way of isolating myself on the ocean for a period of between one and three months (Bombard, 1953, 30)."

Bombard deliberately set out to attract worldwide attention by using a global ocean current as a scientific laboratory. As with *Kon-Tiki*, word of Bombard's plans caused most experts to say that an inflatable boat could never remain afloat on the open ocean more than ten days, which in the event would not matter because Bombard would likely be dead by then anyway. One self-styled expert thought Bombard's choice of a partner for the expedition was fortuitous, since the expert recommended cannibalism if all else failed. "Young man … there is no point in being squeamish. If your companion dies on the way, don't throw him overboard. Eat him (ibid, 50)."

Worse, yet almost equally predictably, the announcement triggered the kinds of waves of charlatans, would-be patrons, publicity-seekers, and unstable volunteers such expeditions universally attract. "One prospective crew-member … wrote to say that he had already tried to commit suicide three times; he asked if he might come with us because he thought I had hit on a workable plan method of achieving his aim (ibid, 36)." Another volunteered his mother-in-law for the proposed voyage. Bombard rejected these in favor of an Englishman turned Panamanian named Jack Palmer. Bombard likewise had to contend with nefarious self-styled sponsors who promised cash and equipment only to find some convenient excuse to back out or disappear when the bills came due.

That is how he found himself in the early summer of 1952, seated with Palmer in a fifteen-foot long rubber boat, ready to leave from the harbor at Monaco, without food, without water, and armed only with his hypothesis about drinking fish water seasoned with seawater. The boat, a very early form of Zodiac inflatable later made famous as part of Cousteau's technological array, Bombard named *L'Hérétique* in order to poke the eye of all who had opposed him.

Bombard was clearly aware of the world-wide sensation caused by *Kon-Tiki*, and that his expedition had the potential to generate similar kinds of attention. He even compared the launch of his inflatable raft, from a French harbor, to that of Heyerdahl's balsa raft from a Peruvian harbor on the other side of the world. It is one of two times he mentions *Kon-Tiki* in his book. Yet the mere fact that it is in the context of an obscure nautical point, and that Bombard does not need to explain his reference, speaks volumes for the fame of Heyerdahl's raft.

"The dinghy was incapable of sailing into the wind. In order to become castaways, we had to start as far away from the coast as possible, as otherwise an unfavourable wind would have driven us straight back again. It was therefore necessary, as with the *Kon-Tiki*, for us to be towed out about a dozen miles off the coast. (ibid, 49).

From Sunday, May 25, 1952, until Wednesday, June 11, *L'Hérétique* drifted westward from the south of France to the Balearic islands off the southern coast of Spain. Hunger pangs had long since become hunger pains, which mutated into a kind of morose lassitude. Neither man had the strength to do much of anything save sleep. Bombard caught a grouper, which provided some of his fish water as well as flesh, which the two managed to choke down without vomiting. As if to reinforce their feelings of being shipwreck castaways, an enormous albino whale, some eighty to one hundred feet long, appeared under the raft, terrifying both men. After surveying them both with menacing red eyes, it disappeared back into the mists from whence it had come, an appropriately Melvillian touch.

When the grouper was gone, the food ration consisted of a glass of seawater and some plankton caught in a net. Emergency rations and water were stowed away where they had been placed under seal upon departure. If the seals were broken, the experiment would be considered a failure. Even so, when an impatient passenger liner encountered the men on Saturday, June 7, Bombard asked for and received a small amount of emergency rations. For this reason, and despite the fact that they had survived on seawater and plankton for ten days, and fish juice and grouper for another four, Bombard's experiment was called a failure.

After landing in Majorca, Bombard sped to Paris to save his reputation. But no potential sponsors would listen to his plea for more equipment. He wanted gear to make ready for a crossing of the Atlantic itself, but now found himself stuck in the Mediterranean. Spurred by *Kon-Tiki*, adventurers of all types and stripes were making ready for sea. "The air was thick with rumours of other expeditions. Someone was going to try paddling from San Sebastian to Dublin in a canoe, and another enthusiast had thought up the idea of crossing the Channel in a 'pedallo' boat... We were laughed at by everyone... (ibid, 88).

Despite his troubles, and leaving his pregnant wife behind, Bombard with Palmer rowed away from Majorca in early July and began a drift toward the coast of southwestern Spain. From there, they shipped as passengers on board a ship bound for Africa and the port city of Tangier. If he could make it that far, Bombard reasoned, then the sweeping transatlantic currents would take him and his raft the rest of the way.

In Tangier, Bombard left Palmer in charge of the raft and its gear while he returned again to Paris, this time to secure replacement for *L'Hérétique*. The raft was three years old and had carried them a thousand miles across the Mediterranean, and Bombard was determined to have a new inflatable raft for the crossing of the Atlantic. The manufacturers, impressed by the performance of the dinghy thus far, in the end provided a new model so the expedition could continue. But the new boat could not rekindle Palmer's old enthusiasm for the expedition. After several false starts from Tangier, during which he was convinced that Palmer was trying to delay a start across the Atlantic, Bombard simply left him behind, and continued on alone.

Teaching himself gradually how to navigate, Bombard attempted to drift and sail beyond the gates of the Mediterranean. He knew he was entering a whole different area of effort, "for when one passes from the Mediterranean to the Atlantic, it is not just a question of rounding a point: a difference of a few miles involves entering another dimension and another age (ibid, 101)." He began to imagine fearful apparitions, at one point mistaking the planet Jupiter for an alien spaceship. Reaching Casablanca on August 20th, he knew that the weeks of drifting on the Mediterranean would now turn to months on the Atlantic. Distances of a hundred miles now became a thousand. Without a companion, loneliness would now attack without mercy.

Bombard remained in Casablanca only four days, before setting out on August 24th on a drift towards the Canary Islands to the southwest. Eleven days later, with his navigation skills, his confidence, and his faith in the raft all strengthened, he landed near Las Palmas in the Canaries. He was now poised to drift alone across more than 2,500 miles of ocean on a small inflatable boat. The French consul and a local French businessman introduced Bombard around the island, and both people and place impressed him. The clientele of the yacht club were particularly notable, as "three quarters of its membership consisted of real yachtsmen and only a quarter of loungers (ibid, 119)."

While in the Canaries his daughter was born, and the newspapers speculated that Bombard was now certain to give up the rest of the expedition. He did return briefly to Paris, where he not only saw his new baby but the courts caught up with him to force him to pay some old fines. After he died crossing the Atlantic, it would be churlish to collect them from his widow and her newly fatherless child. Shoreside delays in both Casablanca and Las Palmas delayed him even further, so that it was not until Sunday, October 19th, 1952, that a French yacht towed Bombard to sea and cast him adrift. The raft set into in the current and its small sail began to catch the northeast trade winds that would push it towards the Caribbean.

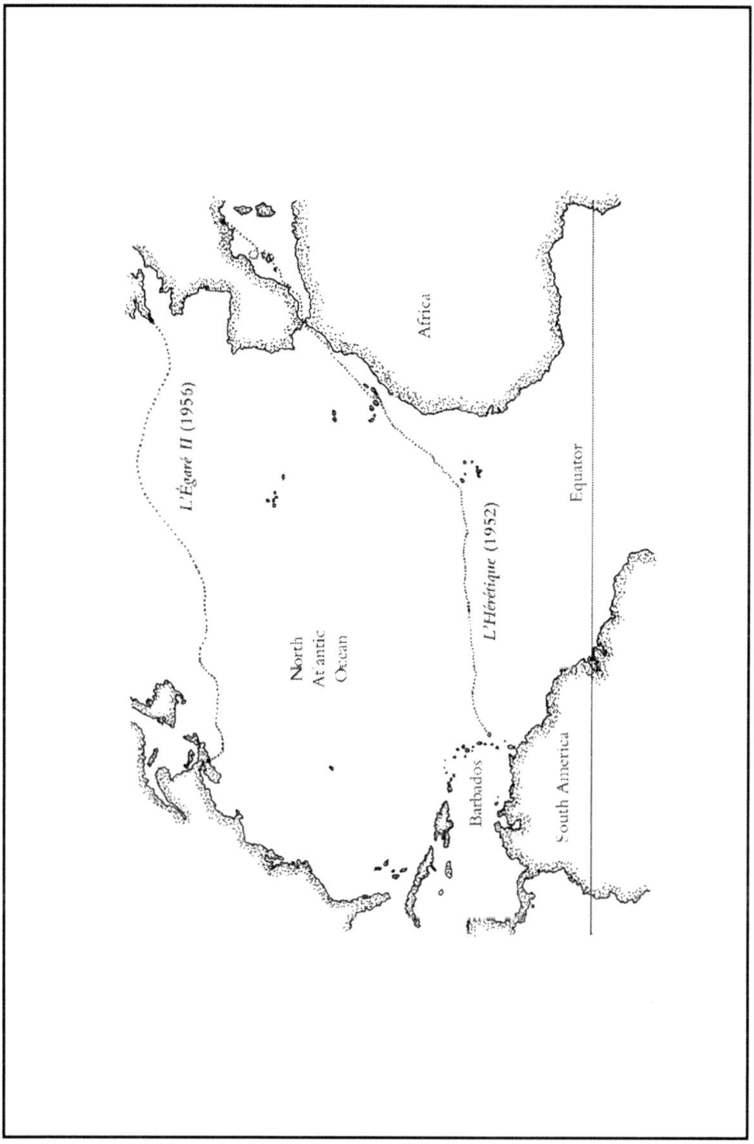

Fig. 4. Routes of the early transatlantic raft voyages (after Bombard 1953; and Beaudout 1957)

Two days out, Bombard discovered that he was not catching nearly enough fish to provide enough water. He drank his usual quantity of seawater, which he thought tasted much less salty than that of the Mediterranean. Dropping off to sleep, he soon found himself in the midst of a hideous nightmare. He imagined himself underwater, with the raft awash, himself drowning. He awoke only to find that he was not dreaming. A massive wave had swamped the raft, and Bombard used first his two hands and then his hat in a futile attempt to bail out his home. Two hours later, the waves subsided and he was able to throw the last of the water overboard, but the effort had exhausted him and soaked all his gear. His only consolation was that, even in the worst weather, his raft had not capsized.

A week from the Canaries, Bombard discovered that, instead of the predictions that he would never catch a fish in the open ocean, his raft had become a small floating shoal followed by clouds of sea creatures. Within a week, he could no longer ascertain his longitude with any accuracy. As a result, he began to believe that he was drifting faster and farther than he really was. On the 28th of October, he broke his watch. Now both his times and his positions became guesswork. He felt dearly the lack of a daily routine, and cursed himself for not devising one beforehand. Enforced idleness from lack of food, combined with the mystery of never knowing what time it was, and the constant cling of cold, damp clothes, became greater adversaries than the ocean.

By the end of October, Bombard thought he was making good enough time to reach one of the Caribbean islands by November 23rd. In reality, he would not arrive until almost the end of the year, and as this truth began to dawn on him, his spirits sank lower. He even lost interest in the books (and musical scores!) he had shipped to retain a hold on his culture. He had brought along "some Molière and a complete Rabelais, a Cervantes, a Nietzsche, Æschylus in the two languages, Spinoza, a selection from

Montaigne and, as musical scores, the two Passions of Bach and the Quartets of Beethoven (ibid, 150)."

As November came and went, with no sight of land, Bombard began to suffer intensely from his diet of fish juice and raw fish. A bothersome shark began poking at the raft until Bombard attached a knife to an oar, sliced the shark open, and left it for the dolphins following the raft. After nearly being capsized at the end of November, Bombard made sure to check that the stock of barbiturates he had prescribed for himself were near at hand. With little possibility of rescue if the raft sank, he was not prepared to endure a day or more in the water awaiting his death.

By Friday, December 5th, Bombard began to think he would never arrive. He was physically exhausted, crippled by diarrhea and persistent body sores. The day previous, his raft was visited by a butterfly. It brought him no comfort, for he now believed that the raft might make land, but hold only a corpse. On the next day, a Saturday, he made out his will.

On Tuesday, December 10th, he sighted a cargo steamer. He had been at sea for fifty-three days without proper food, and the cargo ship informed that he was six hundred miles from where he thought he was. It was enough to force him to quit. But still Bombard refused to give up, though he did accept the offer of a freshwater shower and a meal, his first real food in nearly two months. He reasoned that he could make Barbados in another three weeks. He would arrive just after the New Year. As the cargo ship left him to his raft, he asked the captain if he could find a radio station that would play the Sixth Brandenburg Concerto for him on Christmas Day. It would be, he thought, just enough to see him the rest of the way.

In the event, *L'Hérétique* landed on the island of Barbados on December 22, 1952, with its captain still, if barely, alive. Bombard had demonstrated that a castaway could in fact survive on the most meager of diets from the sea, could in fact survive death by thirst by drinking small quantities of seawater each day. He had lost fifty-five pounds and become anæmic, and discovered that his blood pressure had varied immensely according to his moods. But he was alive.

On Christmas Day, he turned on a radio to hear that the captain of the cargo steamer that had met him offshore had passed a message to the BBC. Bombard leaned back in his bunk to enjoy the Bach concerto he had requested as he drifted on the Atlantic current. He later suggested to boat manufacturers that the bottom of all life rafts should carry printed into the fabric a chart of all of the prevailing winds and currents of all the world's oceans. After all, if he had done it, couldn't anyone?

CHAPTER 3

DE BISSCHOP'S RESPONSE TO KON-TIKI

Science: Tahiti-Nui
Pseudo-science: Tahiti-Nui II
Survival: Tahiti-Nui III

What on earth was that queer craft lying there right in front of the Post Office, in just the same place where our one and only *Kon-Tiki* raft had found sanctuary nine years before?... When I looked more closely I saw ... that the raft at the quay differed from ours in several respects, most notably in being built of bamboo and having two double masts. Bursting with curiosity, I asked my friend the harbor-master whence this new raft came. He gave me a searching look and replied: 'She doesn't come from anywhere, for she was built here. Are you pulling my leg, or do you really mean to say that you have not yet heard of Eric de Bisschop's new expedition? He's going to sail to Chile in that tub—*Tahiti Nui* he calls her—at the beginning of next month, with four other lunatics.

—Bengt Danielsson
From *Raft to Raft* (1960)

Every boat that arrives in Papeete brings stupid individuals who think they have come to the islands of their dreams, where they can live off trees groaning with coconuts, on bananas and fish, and live among "primitive natives" who will be only too happy to provide them with lodging and food, not forgetting the youngest of their daughters. One can understand why the French authorities cast an unfavorable eye on the pitiable victims of [romantic Pacific] literature, human flotsam they will soon be obliged to deport back to their own countries bitter and deceived.

—Eric de Bisschop

Science: *Tahiti Nui* (1956-57)

Thor Heyerdahl's idea of a bearded white god-man bringing civilization to Polynesia from the direction of the rising sun did not sit well with many Pacific islanders. The first direct and sustained Polynesian drifter's challenge to the Norwegian came from a French baron living in Tahiti. By his own admission, Eric de Bisschop had invested the better part of thirty years in a study of Polynesian navigation and anthropology prior to his 1956 voyage in a bamboo raft. But de Bisschop was far more than an armchair explorer. Prior to his raft expedition, a case can be made that only James Cook himself had sailed more of the Pacific Ocean in search of scientific truth than Eric de Bisschop.

Like William Willis, whom we discuss later, Eric de Bisschop was in his sixties, and had already spent much of his life at sea, before living the last years of his life on rafts. In the 1930s, after four years in China, de Bisschop had saved enough to purchase an old Chinese junk, the *Fou-Po*. When a cyclone wrecked this ship off the coast of Taiwan, he built a second, the *Fou-Po II*. On board this second junk, after wrecks in Australia and New Guinea and attacks by wood-eating marine worms called *teredo navalis*, de Bisschop explored whether the 9,000-mile long equatorial counter-current could have served as a prehistoric seaway between Asia and America.

Just before the Second World War, he built a Polynesian double-hulled canoe he named *Kaimiloa*, and sailed it from Hawai'i to France. *Kaimiloa* raced through the first 2,300 miles from Honolulu to the Wallis Islands in little over a month, then crossed the 6,000 miles of the Indian Ocean in less than two. These were extraordinary reaches, and convinced de Bisschop that watercraft of Polynesian design were the equal of any ocean distance anywhere.

Settling in Tahiti after the war, de Bisschop was called back to the sea by the success of *Kon-Tiki*. Two aspects of Heyerdahl's voyage in particular bothered him mightily. First, Heyerdahl's raft, presumably a copy of the pre-Columbian sailing raft, was built with fixed centerboards. De Bisschop knew this not to be the case: these centerboards had been designed to be moved up and down according to the navigational needs of the raft, a fact Heyerdahl himself learned in 1952, five years after his first raft trip. Similar centerboards are used to control the movements of bamboo rafts of ancient design in Taiwan and on the coasts of China and Viet Nam.

These centerboards, or guaras, allowed one to tack and cross a raft into the wind like any European sailing vessel. As Heyerdahl wrote in 1994 to my colleague John Haslett, then planning the first of his three balsa raft voyages, "balsa rafts of from 3 to 5 balsa logs are still used in several fishing ports in both Ecuador and north Peru, and they go out at night and come back to the same beach by noon" (Heyerdahl 1994). Long before Heyerdahl's experiments with the guaras, Eric de Bisschop met *Kon-Tiki* crewmember and anthropologist Bengt Danielsson in Tahiti after the expedition, and Danielsson had to admit that the centerboards had likely not been employed properly.

Much more than this esoteric question of raft design, however, was the suggestion that Polynesians could not have reached the shores of South America because the prevailing winds and currents would have effectively prevented such a passage. This de Bisschop refused to accept. He himself had tacked Chinese junks and double-hulled canoes against the prevailing winds for much of his life, and saw no reason why prehistoric Polynesians could not have accomplished the same thing. What was startling, however, was the technology de Bisschop chose to employ in an attempt to refute the *Kon-Tiki* theory.

When Bengt Danielsson arrived in Tahiti for the fifth time in the fall of 1956, he saw a raft moored along the docks of Papeete in the same spot where he and the rest of Heyerdahl's crew had stepped off *Kon-Tiki* almost ten years earlier. But this was a raft of a very different sort. Again like William Willis before him, de Bisschop had distilled a lifetime of sea-going experiences into the design of a transoceanic raft. In de Bisschop's case, his raft reflected his experiences in Polynesia and China, as well as his knowledge of Chinese and Peruvian centerboards.

The raft was built of bamboo, equipped with Peruvian guaras, and rigged like a double-masted Chinese junk. Remarkably, rather than use all his accumulated experience to demonstrate that a true Polynesian double-hulled voyaging canoe could manage a journey from Tahiti to South America, de Bisschop proposed that his polyglot bamboo raft could make the same voyage against five thousand miles of prevailing winds and currents. He advanced his proposal to general disbelief. When he announced that he would set his course far to the south, to take advantage of the shifting westerly winds and currents that prevail around 40° south, most sailors thought the expedition a suicide. Any raft caught in the cold and "roaring forties," they argued, would be quickly torn to pieces.

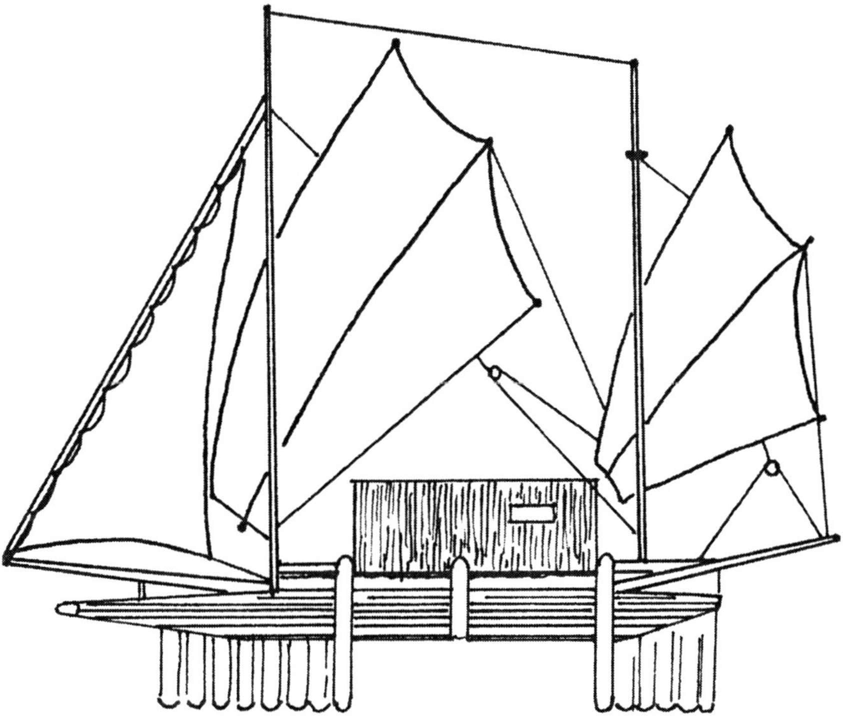

Fig. 5. De Bisshop's *Tahiti-Nui I* (after Danielsson 1960)

To de Bisschop's credit, he was merely taking anthropological theorizing to its logical conclusion. Like prevailing winds, the prevailing anthropological view was one of an ultimate Southeast Asian origin for Polynesian culture. De Bisschop on the other hand, considered the "Polynesian problem," as it was often referred to, far from solved. Like Heyerdahl, he had no time for academics with no practical maritime experience upon which to base their theories of Polynesian origins and migrations. Heyerdahl had had the courage to put his raft where his theory was. For this reason, and this reason alone, de Bisschop told Bengt Danielsson that he had more respect for Thor Heyerdahl than all his other opponents put together.

In this, as in other notable aspects, de Bisschop was far ahead of his times. In de Bisschop's view, if anthropologists wanted to understand the essentially maritime culture of the Polynesians, they had better start to understand the nature of the maritime world. That meant a vast expansion of scholarly geographic horizons, and a total abandoning of the idea (still subconsciously prevalent in many academic discussions) that the geography of the Pacific has remained unchanged since time immemorial. De Bisschop railed against anthropologists who "blandly assume that the geographical features of the Pacific and Indian Oceans, with all the lands which emerge from them or border them, have not budged an inch for thousands of years—an assumption based on nothing except perhaps man's subconscious reluctance to admit that he inhabits an unstable and ever-changing crust." (de Bisschop 1959, 7)

De Bisschop saw a Polynesian sphere of influence extending from Easter Island and perhaps the shores of South America in the east, to Madagascar off the coast of Africa in the west, a span more than half the distance around the world. No other peoples could claim such enormous dispersions. For de Bisschop, no outsiders had taught Polynesians to sail, rather it was the Polynesians themselves who, by spreading their seagoing knowledge from Indonesia and India and then all the way to Madagascar, had accomplished quite the reverse. He imagined the Polynesians visiting South America before the time of Christ, to return with plants common to both places. And this nearly two thousand years before Europeans made tentative voyages with the currents and winds from Spain to the Caribbean. With such a tradition, de Bisschop proposed rewriting the maritime history of the world, with the Polynesians in their rightful place at the center of that history.

To study this enormous problem, de Bisschop envisioned as well a new field of maritime ethnology, where scholars would use re-created voyaging technologies in order to study ancient diffusions. That genetic evidence now points to the likelihood of transoceanic migrations to Madagascar originating in the Pacific, and that experiments in long-distance voyages by recreated Polynesian canoes are now integral facets of maritime anthropology, only reinforces de Bisschop's pioneering and little appreciated role in Pacific studies. Like Heyerdahl before him, de Bisschop believed that no attention would be offered to his ideas unless his took them to sea, on an actual transoceanic voyage. His bamboo raft would be the instrument of his attack on academic convention.

Like Danielsson, many wondered why de Bisschop would choose a raft for his experiment, rather than the double-hulled and double-outrigger canoes he was so familiar with. De Bisschop was clearly stung by this repeated criticism. He countered that, in his studies of Polynesian navigation, a maritime culture as sophisticated as that of the Polynesians, surely would have possessed different vessels for different missions. He imagined that single outrigger canoes were appropriate for skimming over shoals and reefs; much larger double-hulled and double-outrigger canoes would have been used for quick hit-and-run raids on neighboring tribes or voyages to known destinations.

But for long-distance voyages of exploration and colonization, with their essential cargoes of people and provisions, only large rafts would suffice. And as de Bisschop saw it, when equipped with the moveable centerboards called guaras, these rafts became sailing vessels capable of true navigation. "It was by no means a floating contraption at the mercy of winds and currents" (de Bisschop 1959, 34-35). Furthermore, the slow speed of a raft voyage, de Bisschop imagined, was no hindrance for a Polynesian people with a fundamentally different view of time.

To test whether Tahitian bamboo could remain buoyant for the duration of a long ocean voyage, de Bisschop had a diving platform constructed from bamboo and moored near a friend's house for a year. While some of the bamboo was eventually attacked by *teredo navalis*, the platform was still afloat after a year. As for whether or not Polynesians ever used bamboo rafts, de

Bisschop was unequivocal. He cites several voyages on bamboo rafts navigated by prehistoric Marquesans fleeing tribal wars. Some of these bamboo rafts, constructed with five layers of bamboo logs, ventured as far as Tahiti and even Hawai'i.

In building the raft, de Bisschop had little to go on in the way of local knowledge. If bamboo rafts had ever been used in Polynesia, they had long since become distant memories. So he relied instead on his knowledge of sailing, on the assumption that, confronted with similar building materials on the same ocean, ancient Polynesians would have come to the same nautical design conclusions. Retreating to the solitude of the island of Rurutu, where he had a home and a wife, de Bisschop built a one-tenth scale model of the actual raft he hoped to build in Tahiti.

> What a joy it was to come back to the pleasant fare (home) which I had built facing the sea, to my sweet vahiné, and above all to the peace and silence of the nights, broken only by the muffled booming of the surf pounding on the reef—a peace so light that you could feel it floating everywhere, only waiting for the slightest opening of a door to insinuate itself into the most secret fibers of your heart and soul. (de Bisschop 1959, 57).

Returning to Tahiti with his model, de Bisschop set up a quarantined area at the naval dockyard where he could build his raft and at the same time "keep out the inquisitive and the inevitable 'advisors'" (ibid, 60-61). Among those inquisitive visitors peering over the fence at de Bisschop's raft workshop were two of the world's greatest explorers, one already famous and one whose fame was still twenty years away. Thor Heyerdahl, returning home after his intensive and pioneering stratigraphic archaeological Easter Island excavations, "shook his head seriously and solemnly declared that he would never dare entrust his life to such a fantastic craft" (Danielsson 1960, 61).

The second visitor, a fascinated graduate student by the name of Ben Finney, hesitated to disturb the work going on behind the secured naval compound, and so lost his only chance to meet de Bisschop personally. But Finney would later revolutionize the world's thinking about the seagoing capacity of the Polynesian double-hull canoe. With Heyerdahl, Bengt Danielsson, Finney, and de Bisschop and his crew all on Tahiti at the same time, was a young American sailor by the name of Norman Baker, who later in life would become the navigator and second-in-command of Heyerdahl's three reed boat expeditions. Never before and likely never again would there be assembled in one place so many explorers who already possessed or would later amass such an overwhelming amount of experience in drifting and sailing prehistoric rafts and canoes.

Those who did dare venture through the gates of de Bisschop's compound were the usual assortment of would-be adventure-seekers and potential crewmembers. As Bombard had so sharply discovered, and de Bisschop himself sarcastically noted, these drifters were individuals who were "keen enough at first, but who, as the hour of actual departure approaches, gradually deflate like a tire with a slow leak" (de Bisschop 1959, 68).

When the bamboo was ready, it was lashed together with coconut fibre rope. The sails de Bisschop had plaited from vegetable fibres. Once finished, he christened the whole creation Tahiti Nui, Great Tahiti. The one concession to the twentieth century was a cabin made of double walls of plywood, which housed an echo sounder, radio, a darkroom, and a dry sleeping area for the crew. Unlike Bombard, whose voyage de Bisschop cited, he had no desire to conduct a human endurance test, "to be made to swallow plankton and other revolting stuff of that sort, to drink sea water or the juice squeezed out of raw fish" (ibid, 109). De Bisschop himself was sixty-five years old and debilitated by bronchitis and emphysema. His doctor told him in no uncertain terms that he would not survive the journey, a knowledge that bothered de Bisschop not at all. He knew his time was close. He felt death "like a patient, comfortable, smiling friend, near at hand, only waiting your pleasure to fall into step with him to make the long journey toward the Unknown" (ibid, 79).

By de Bisschop's reckoning, the last great voyages by a Polynesian fleet had taken place some 700 years earlier, during the fourteenth century. On November 8, 1956, he prepared to follow that fleet to sea, on a strange polyglot raft at the start of an even stranger scientific experiment. De Bisschop reckoned that the voyage to South America would take between three to four months. Provisions were stowed for five, "beer and lemonade … a dozen enormous stems of bananas, numerous sacks of potatoes, kumara, onions, taporo, gourds, not to mention mountains of coconuts both green and dried" (ibid, 105). Any longer than that on a raft and de Bisschop feared the consequences. "It is not only bamboo that begins to degenerate after seven months at sea," he wrote laconically (ibid, 93).

Like all previous raft expeditions, *Tahiti-Nui* began with a tow by a diesel-powered ship to its place of departure. Fifty outrigger canoes escorted the raft from Papeete harbor, perhaps unconsciously suggesting a method by which prehistoric rafts were maneuvered off the beach and into position to take advantage of wind, tide, and current. Chased by a friend's yacht, the crew enjoyed one last view of Tahiti in the form of a beautifully tanned young woman who, as de Bisschop warmly related "lifted a corner of her bikini toward us (I cannot say more!) in a charmingly simple gesture of adieu. I turned to my crew and said, "Take a good look, my boys—it will be many moons before you see anything like that again!" (ibid,102).

In the event, only four moons passed. Even before clearing Tahiti, de Bisschop became concerned with the raft's buoyancy, and decided to put ashore to lay in additional bamboos for the long voyage ahead. The same gunboat that had towed them out, now returned to tow them back in. To Alain Brun, de Bisschop's second-in-command, the gunboat saved them all from embarrassment when it "mercifully took us to a remote creek on the south coast of Tahiti" (Danielsson 1960, 63). Thus reinforced, the raft started south toward the Austral islands of Rurutu, Raivavae, Tubuai, Rimatara, and Rapa, making three and a half to four knots of speed on a following wind that lasted for a week.

At this early stage of the experiment, the raft surprised even de Bisschop by its ability to use its guara centerboards to make a passage toward the southeast. Before the voyage, de Bisschop thought the best the raft might do was approach his own Austral island of Rurutu. Now he found himself approaching the seas that separated Raivavae and Tubuai, more than 200 miles east of Rurutu.

Heyerdahl had just called at many of these islands after his work on Easter Island. William Mulloy had carried out extensive archaeological excavations at the spectacular aerie at Morongo Uta, one of twelve entirely unexplored mountaintop fortresses on Rapa Iti. Arne Skjølsvold, the Norwegian archaeologist who had discovered the inscrutable "kneeling statue" of Easter Island, spent several weeks mapping the ceremonial *marae* platforms of Raivavae. Contrary winds now blew the *Tahiti-Nui* in a complete circle around Raivavae. A few weeks later the raft passed the latitude of Rapa Iti, and so moved beyond the limits of French Polynesia. As 1956 turned to 1957, 5,000 miles of cold open sea lay between the raft and its destination in Chile.

Beyond Rapa, at about 33° south, the raft picked up shifting west winds and de Bisschop set his course directly eastward to South America. For two months, *Tahiti-Nui* careened furtively eastward in tolerable temperatures that hovered between 68°-77° F. On February 23, 1957, the raft passed the longitude of 117° W, the halfway point on its voyage. But the mark brought little consolation. The experiment was now three and a half months old, at a point when de Bisschop had believed they would be safely ashore in Chile. Instead, the bamboo, put in the water in September, had been afloat now for more than five months. It was approaching the limits of its buoyancy with still more than 2,500 miles to go. The crew, to the contrary, thought the worst was over, that the remaining miles would speed by.

Instead, they were met almost immediately by a dreadful fortnight of winds blowing from the east. De Bisschop had told Bengt Danielsson that he intended to sail down to 40° south, where he would be assured of steady winds and currents from the west. But now he hesitated, staying in an area of wavering winds around 35° south. Even here the seas were rough, and de Bisschop was convinced *Tahiti-Nui* would lose its two masts if he tried to sail any farther south.

Ben Finney, for one, believes it was the only decision de Bisschop could have made. At 40° south, the raft would have been torn apart by mountainous seas. If prehistoric Polynesians had voyaged along this route to the east, they did so only at great risk to themselves and their expeditions. Even at 35° south, heavy winds forced de Bisschop to take in most of the sail the raft carried, to prevent it from being carried away.

When the raft began show signs of breaking up in late February, de Bisschop was forced to put down a minor mutiny by the three other crewmembers. All were half de Bisschop's age and cared little for the scientific substance of his experiment. They advocated instead an audacious retreat as far as 50° south in an attempt to speed their passage. Daily radio interruptions suggesting steady winds further south—and thereby further enticing the demoralized crew in that deadly direction—nearly drove de Bisschop to pitch the set overboard. When the raft circumscribed a complete circle on March 11, returning the crew to a point they had passed seventeen days earlier, morale sank even further.

One unexpected advantage of de Bisschop's more northerly course was a near miss of Easter Island during the first week of March. The seagoing raft wandered to within 350 miles of anthropology's most enigmatic island on March 7, 1957, demonstrating a plausible access route from the west to this most remote corner of Polynesia. De Bisschop in fact began to spin a theory that the island had been originally colonized by a raft caught in the same contrary winds his raft now endured. His crew wished he would follow his own hypothesis and make for a landing at Easter Island, at the very least so that repairs could be made to the raft.

Had de Bisschop done so, it is likely that the voyage of *Tahiti-Nui* would have taken a rightful place alongside *Kon-Tiki* as one of the great Pacific drift voyages, and ironically increased its value to maritime anthropology. *Tahiti-Nui* had linked the Austral islands of Rapa Iti and Raivavae, with their stone fortifications and ceremonial marae platforms, with the stone ahu platforms and carvings of Easter Island. The raft had voyaged almost as far as *Kon-Tiki*, and remained afloat for more than six months, four of them on the high seas. By continuing eastwards to certain destruction, de Bisschop weakened the plausible case he had already made for the efficiency of the long-distance seagoing bamboo raft in prehistoric Pacific expeditions.

That destruction arrived in slow and painful measures. The raft drifted through April as the crew suffered through a near-total lack of fresh water. May brought with it fresh winds from the west, but the raft was still a thousand miles from Chile. A week later, still eight hundred miles from the coast, the big four-inch main bamboo logs began to break away in fifty mile an hour winds. The situation now became desperate, with the raft listing heavily. To worsen matters, de Bisschop found his bamboo hull "riddled with tunnels the size of your little finger, each one with its fat white [*teredo navalis*] worm."

> They have wicked heads with two hard curved plates at the business end, only too well designed for the dastardly work of boring and destroying.

> I have seen natives, especially in Melanesia, reveling in these large white worms, which they eat raw. Here, now, is a field of survey which has been ignored by the specialists—something to add to the menu of those who cast away at sea. I myself had never thought of it. I wish I had; it might have been most useful when, on *Fou Po II*, I went for nearly three weeks [without food]. But how I could have harvested the little beasts into the frying pan when they were snug below the waterline in the very planks of the hull which kept the boat afloat, I don't know. (ibid. 185-86)

By the middle of May, after six months at sea, even de Bisschop was tired of the cold southern seas, and began to long for the warm and light blue waters of Polynesia. When a severe storm forced the crew to abandon the idea of a landing at the Juan Fernandez islands off the coast of Chile, where Alexander Selkirk had found himself marooned in 1703, de Bisschop at last signaled for a tow. Over their radio they listened to spurious reports that the raft had been dismasted, that the crew was injured by the attack of a giant fish, that giant mollusks had attached themselves to the raft and were dragging it down. The expedition had taken on a distinctly Vernian tone.

The raft plowed on to its farthest point east: 87° 54'W. Winds then forced it back more than a degree to the west, where a Chilean naval vessel caught up with it on May 22, 1957. The unsuccessful tow resulted in the final break-up and abandoning of the raft on May 26th. As the starboard tiki god figure, carved by a Marquesan artist, was salvaged from the raft, de Bisschop heard the final sickening splintering of the bamboo. On board the Chilean ship, de Bisschop took to a bunk and lamented his failure to prove his theory.

Once ashore in Chile, de Bisschop began writing up his experiences on board the *Tahiti-Nui*. Now, at the age of sixty-six, he was more determined than ever to complete a full-circle voyage from Tahiti to South America and back. On February 15, 1958, a new raft, christened *Tahiti-Nui II*, was prepared to drift from South America to Polynesia, along the track pioneered by the balsa rafts of Thor Heyerdahl and William Willis. It would prove to be the final expedition in the long, adventurous life of Eric de Bisschop.

Pseudo-Science: *Tahiti-Nui II* (1958)

> I've had my fill of knocking about the world and the oceans, and I can only thank the gods for having made life so wonderful for me—so colored, so varied. I could depart this life today happily and with no regrets, and what better way for a sailor to die than by making a hole in the sea?—though I would have preferred a somewhat gayer setting for my last plunge, a bluer sky and a more inviting, more warmly embracing sea.
>
> —Eric de Bisschop
> *Tahiti Nui* (1959)

Drifting toward his death in the mid-Pacific in 1958, Eric de Bisschop made once last call upon the Polynesian god of the sea. He had always believed that he would die somewhere in the Marquesas archipelago, where Latitude 10° South meets Longitude 140° West, and now it seemed as if his premonition was a true one. He had always been drawn to the Marquesas, a chain of islands settled by Polynesians around the time of Christ.

The Spaniard Mendaña landed in the islands in 1595. James Cook explored the area for a few days in 1774, as did Thor Heyerdahl in 1956. The *Essex*, a U.S. Navy vessel under the command of David Porter, launched a seven-week war against the native Taipi in 1813, one of the very first American actions against native peoples outside North America. It is not an easy place to fight a naval action, with its 1,000-foot high cliffs.

The famous Stonington captain, Edmund Fanning, arrived at Nuku Hiva in the Marquesas in May of 1798 on one of his five circumnavigations of the globe. Commencing the usual trade with the natives, he observed "something wrapped up in palm leaves, on board one of [the canoes], the native in which did not offer it for barter." Fanning examined the item, finding to his surprise that it was "a piece of human flesh, baked." The individual explained that it was a part of one of his enemies, and so of course very good to eat, and that "whenever he was hungry, he was going to eat it."

The crew of twenty on board a Nantucket whaleship also called Essex could have sailed south to the Marquesas were they unafraid of cannibals; or northwest to Hawai'i, were it not for hurricanes. Operating in the Pacific on November 20, 1820, the whaler was rammed twice and sunk by its supposed prey, an eighty-five foot long sperm whale. There began one of the most agonizing sagas of the sea, the basis of Melville's *Moby-Dick*, and a story so bizarre that even Melville left out the gruesome details when he wrote the greatest of all sea novels. Rather than make for the cannibals of the Marquesas, Captain Pollard of the Essex headed his surviving boats southeast, even though they were over three thousand miles from the nearest land in that direction, and against wind and current for good measure.

The boats from the *Essex* arrived on Henderson Island in December, where three crewmembers decided to remain and try their luck at survival on the minute island. Captain Pollard himself insisted on heading for South America, and the remaining crew sailed dutifully on, only to endure storms, sharks, starvation, thirst, madness and, ironically, cannibalism. Three crewmembers were lost when one of the boats disappeared in December. Two others died and were buried at sea. By February of 1821, with the men starving after two and a half months in open boats, the Nantucketers were forced into the kinds of flesh-eating that so repulsed them when practiced by Marquesan natives. As the Nantucketers died, they were eaten.

First mate Owen Chase wrote later of the fate of one of the dead men: "We separated his limbs from his body, and cut all the flesh from the bones, after which, we opened up the body [and] took out the heart ... which we eagerly devoured." On board Captain Pollard's boat, cabin boy Owen Coffin was not so fortunate. The four survivors drew lots to choose who among them would die so that the rest might live. When Coffin drew the morose assignment, Captain Pollard remarked as gently as possible given the circumstances, "My lad, if you don't like your lot, I'll shoot the first man that touches you." To which the cabin boy replied, "I like it as well as any other." Whereupon his starving mates quickly dispatched and devoured him. Only five from the Essex were picked up at sea later in February, surviving to testify to what quickly became a Pacific legend.

Native Marquesan navigators, with their specialized local knowledge, sailed from the northwestern islands to their southeastern neighbors during November and December, when the dominant winds from the east slackened somewhat, a technique rediscovered by anthropologists centuries later. Their business concluded, they would then wait for the prevailing west winds to blow them home.

But now, as Eric de Bisschop's raft built of Chilean cypress sank inexorably toward the abyss, those same prevailing winds and currents pushed *Tahiti-Nui II* north of the northernmost island in the Marquesan chain. As a mariner who had circled most of the earth in a long life of adventure, de Bisschop knew all too well what this meant. His raft expedition might drift into the wastes of the central Pacific, where islands were as scarce as water on the moon. He and his four companions would die miserably, of thirst or madness. It would be the *Essex* all over again.

Thirst had already driven one of de Bisschop's crewmen crazy. He had begun to literally saw the raft apart, in order apparently to build himself a private escape pod. De Bisschop, rather than exhibit the caustic temperament he showed to many during his life of adventure, treated the wretch with a gentleness most had never seen him display in far less strenuous circumstances.

After the sinking of the bamboo raft *Tahiti-Nui*, de Bisschop intended to build a copy from the same materials. After all, the original had floated for almost nine months, and even though it missed the Chilean coast by 800 miles, this was a creditable showing after a voyage of more than 4,000 miles from Polynesia. But no bamboo was available in Chile's cold climate, nor did balsa trees grow in the countryside.

When a collection of shipyards in the port of Constitucion offered to replace his bamboo raft with a small cutter—and attempt to drum up some Polynesia cutter business into the bargain—de Bisschop protested that he required a raft to continue his ethnological studies. The shipyard owners replied that their only experience in rafts involved a large oak car ferry ordered by the government, which had promptly sunk upon launch. A compromise was reached in the fall of 1957 wherein the shipyards would build a raft of approximately the same dimensions of *Tahiti-Nui* from buoyant Chilean cypress wood. De Bisschop then retreated to a friend's home to write a book of his experiences on the voyage from Tahiti, and left the work of building what would become *Tahiti-Nui II* to Alain Brun.

Unable to lash the cypress logs together, Brun fastened them instead with hardwood pegs. The principle of navigation would be similar, however, as Brun rigged the raft in the Chinese junk style of the first raft, and added several guara centerboards as well. When de Bisschop arrived back in Constitucion in January of 1958, with his manuscript complete, his raft was ready as well. A month later, with four young crewmembers including Alain Brun, the raft put to sea. De Bisschop chose to cut across the Humboldt Current and sail offshore as far as Callao in Peru, 1,500 miles to the north, where the raft would be pointed toward Polynesia. He adapted himself immediately to a renewal of his drifter's life:

> [Eric] simply hung up his briefcase on a nail over his bunk and took out his beloved papers and books: then he felt at home at once. This matchless power of adapting himself was

> of course due mainly to the fact that he lived entirely in a private world of his own and seldom noticed where he was and what was going on round him. (Danielsson 1960, 117)

The raft proved extremely simple to handle, but not so the human relationships on board. Almost immediately, the two newcomers to the crew skirmished with de Bisschop—a difficult personality in the best of times—and a pattern of shifting alliances began that would ultimately doom the expedition.

The first half of the coastwise journey to Callao took three weeks, at an average of forty miles per day. At this speed—about the pace of a leisurely walk—*Tahiti-Nui II* slowly drifted past many of archaeology's most intensely interesting prehistoric coastal sites. On March 21st, 1958, the raft maneuvered along a "steep stony beach, inside which a desolate sand desert stretched as far as we could see in every direction" (ibid, 126). The expedition had reached the land of the ancient Nazca. These people, creators of the famous "geoglyphs" that ornament this coastal desert, populated the southern coast of Peru from about the time of Christ to the middle of the sixth century. It is the same approximate time horizon as the Moche people of Peru's north coast, the likeliest candidates for the balsa raft navigators that Thor Heyerdahl believes drifted into the eastern Pacific sometime around 500 C.E.

Three days later, the raft passed the prehistoric burial grounds of Paracas, a site discovered and excavated by the first great native Peruvian archaeologist, Julio C. Tello. The burial grounds were in use during the five centuries or so prior to the Nazca and Moche periods of Peruvian coastal prehistory, and contained both cavern-type and masonry vault-type crypts. These crypts, with their mummified human remains, attest to a belief dating to least 4,000 B.C.E. in Peruvian prehistory that human remains be salted and preserved, for only a physically intact corpse could "enter the afterlife and join the realm of the ancestors" (Moseley 1992, 94).

Fig. 6. De Bisshop's *Tahiti-Nui II* (after Danielsson 1960)

Thor Heyerdahl noted that several hardwood guara centerboards have also been excavated from this area, and dated to the same period as the Paracas burials. The ocean just offshore from this necropolis was an area of shifting currents, and forced the raft's crew to keep constant watch for, as Alain Brun noted, they "had no desire to end our days in this burial ground, however famous it might be…" (Danielsson 1960, 128).

On March 26th, the raft was taken in tow by a Peruvian naval vessel, and pulled the final ten miles of the distance to Callao. This was accomplished with such skill that Brun laconically surmised that "Peruvian patrol boats were accustomed to towing rafts of prehistoric model" (ibid, 129). An inspection of the raft revealed no teredo infestation of the cypress logs, which appeared to still possess about ninety percent of their original buoyancy. Even so, de Bisschop thought it prudent to reinforce the raft with a dozen balsa logs obtained free from local benefactors, as well as fourteen aluminum water tanks bought from a Belgian chemist living in Lima. Unbeknownst to the crew at the time, these tanks would later form the basis of *Tahiti-Nui III*, the rescue raft that would save all but one of their lives.

During their stop in Callao, Eduardo Ingris, captain of a balsa raft during a misbegotten voyage from northern Peru to the Galapagos in 1955, the year after Willis' balsa voyage to American Samoa, begged de Bisschop to be allowed to join his crew. Even with the early splits in his expedition, de Bisschop resisted Ingris' appeals, and with his original crew was towed from Callao into the Humboldt Current on Sunday morning, April 13th, 1958. It was eleven years since *Kon-Tiki* had drifted away from the port and into history.

To avoid Ingris' fate on the *La Cantuta* drift, de Bisschop steered a course to the west, across the currents that would sweep them toward Panama if the raft were allowed to drift uncontrolled. One week out from Callao the raft, like a train switching tracks, left the north-setting Humboldt Current and picked up the South

Equatorial Current, which set the raft on a course for the mid-Pacific. The voyage at this point was so easy that Brun occupied watch after watch discussing navigation and anthropology with de Bisschop. Another of the crew took to comparing the daily speeds and distances of *Tahiti-Nui II* with *Kon-Tiki*, having brought a copy of the German edition of Heyerdahl's classic book on board the raft. As Brun considered this most "original race with eleven years' difference in time," *Tahiti-Nui II* was holding her own against her balsa raft predecessor.

"As everyone knows, the *Kon-Tiki* men wanted to prove that it was possible to sail from South America to Polynesia on a balsa raft, but it really made no difference to them at which of the many Polynesian islands they ended their voyage. We, on the other hand, were firmly resolved to reach Tahiti, and therefore kept a rather more northerly course than our predecessors to avoid the insidious coral reefs of the Tuamotu group. But this did not prevent us from comparing their daily runs with our own. We soon found to our satisfaction that we were keeping up at least as good a speed as the *Kon-Tiki* raft, i.e., thirty-five to fifty miles a day…" (Danielsson 1960, 139).

The ease of the first few weeks dissipated as Brun found that the crew had trouble coping with the many steering guaras: six forward and eight aft, that controlled the movement of the raft. There simply were not enough crewmembers to shift the guaras rapidly enough when the situation called for such maneuvers. Yet the first half of the expedition to Tahiti was accomplished in just six weeks, and on June 2, 1958, the crew found that the raft had drifted eighty miles during the previous twenty-four hours, their best run yet.

When the rudderpost snapped, three of the forward guaras were pulled up and nailed together to form a new and larger rudder. When the new rudder required more muscle to keep in trim, the crew pulled up even more guaras to build a platform for the helmsman to stand upon to avoid being washed overboard. The larger rudder was needed immediately. The more northerly course de Bisschop had steered now began to backfire, as the raft stubbornly refused to turn south toward Tahiti. They had assumed that, like *Kon-Tiki*, their raft would begin its southerly arc toward Polynesia when it passed 120° west. Yet as the raft passed that longitude it continued to drift west at around 4° south, and as they all knew, Tahiti itself lay at 17° south. To make matters worse, they discovered that *Tahiti-Nui II* was slowly sinking out from under them. In their repeated comparisons with *Kon-Tiki*, the crew found the *Tahiti-Nui II* wanting in both trajectory and buoyancy.

By the middle of June, eight inches of water was sloshing over the decks and filling the cabin. The crew moved their sleeping bags onto the flat roof of the cabin, as de Bisschop himself began to grow weaker and feel the chill of the night winds even more than the younger men. By June 20th, it became clear that the trajectory of the expedition was taking them away from their Tahitian destination. A landing in the Marquesas islands, still four hundred miles away to the west, was still possible if they were favored with sustained northerly winds. But, perversely, as soon as they wished for them, the wind veered over to the southeast, blowing the raft further north, away from the Marquesas.

During the midnight watch on the evening of June 26-27, Brun watched the raft suddenly lose its buoyancy and sink three feet. He left the rest of the crew to their sleep during the emergency, but in the morning they examined the cypress logs to find them riddled through by *teredo navalis* worms. Brun cut a piece of one log and threw it into the sea. It sank immediately. With the raft awash and increasingly difficult to steer, most of the crew wondered what the point was of keeping continued watches. The raft was now unable to make steerage way, and had become instead a derelict drifting at the caprice of wind and current.

To Brun's surprise, de Bisschop agreed with the crew, telling the men to save their strength and let the raft drift where she liked. As the raft drifted far north of the Marquesas on July 1st, 1958, a helpless hulk, de Bisschop called upon Taaroa, the Polynesian god of the sea, to look after them. Brun, refusing to submit to the fates, insisted that the men maintain some semblance of naval discipline or risk drifting benumbed into the wastes of the central Pacific. De Bisschop, telling the men that he was too "old, tired and ill" (ibid, 154) to force his will upon them, turned command of the raft over to Brun, and retired to his sleeping box on the roof of the cabin. De Bisschop was no doubt thirsty as well, as only fifteen of the 100 gallons remained of water the raft had started with. Worse, since leaving Peru, they had experienced not a drop of rain. The daily ration of freshwater was now mixed with increasing quantities of seawater.

When two of the crew wanted to build a small boat and try to row back to Marquesas, Brun and de Bisschop managed to dissuade them. Instead, the main masts were cut away and replaced by a smaller mast carrying a tiny storm sail. Yet the raft continued to drift further to the west-northwest, into the most desolate area of the Pacific. Morale diminished further in mid-July when then crew sighted a cargo ship and the vessel passed within three miles of the raft without seeing it.

As July turned to August, the crew was in outright mutiny, openly laying plans to build an escape raft. Brun squelched this uprising by pointing out that they were 300 miles from Caroline Island, and moreover were drifting to the north of it. Only if the *Tahiti-Nui II* managed to drift within close proximity of the windward side of an island, would they take the very great risk of constructing an escape raft and attempting a landing over a coral reef. At this point, Eric de Bisschop made it clear that he was not leaving the raft under any circumstances. This did nothing to quiet the malcontents, one of whom had evidently gone mad from thirst and hopelessness. When de Bisschop tried to calm the man, he was threatened with an axe. De Bisschop and the rest of the crew then held a ship's council, at the conclusion of which they swore out a document allowing the madman to build his escape raft, provision it with his share of stores, and sail away.

The night following the mutiny, the first rain since the raft left Peru began to fall, and the spirits of the crew rose a bit. The crewmember building his escape pod regained his senses, and the expedition members gloried in both a renewed supply of freshwater and a bottle of brandy. Yet the same rain that saved the humans now made the cypress raft even more top-heavy. With their options rapidly running out, the crew tried to steer towards one of the copra islands of Vostok or Flint, where William Willis had passed so closely by on his balsa raft only four years earlier, in far the better health and spirits. But unlike Willis' beautifully designed sailing raft, the *Tahiti-Nui II* was now a useless derelict, rolling heavily with each storm. The life of the crew was reduced to a ten-foot by thirteen-foot cabin roof.

On July 29, 1958, Brun took a sun sight that placed the raft on a course to miss all of the copra islands by a wide margin. They still had a chance of reaching Starbuck Island, a small, uninhabited atoll. But even Starbuck was 400 miles away, and the raft by now was completely unmanageable, listing along at a meager twenty-five miles a day. By August 6th, still 250 miles from Starbuck, the crew had taken to throwing all extraneous weight overboard, including all of the scientific water samples one of the crew had made all the way from Callao. That night the crew decided it was time to build their escape raft. If they approached Starbuck or any other island, they would make a break for shore and salvation. Unbeknownst to the men, the only things that would meet them at Starbuck would be a thousand-yard-wide coral reef, the ruins of a British guano mine from the 1880s, and a complete absence of any fresh water.

The *Tahiti-Nui III* took shape over the course of the next four days, as the crew frantically tried to ready their rescue raft before the *Tahiti-Nui II* drifted past their rendezvous with Starbuck Island. They were like astronauts gauging their reentry into earth's atmosphere, and the slightest miscalculation would send them back into a hopeless void of ocean. The aluminum water tanks they had almost scorned in Callao, now came back to save their lives. Yet it was no small feat to create a small escape raft from the hulk of another, larger, sinking raft, with both bobbing in the waves, one threatening to sink at any moment while the other sought to sail away.

Survival: ***Tahiti-Nui III*** **(1958)**

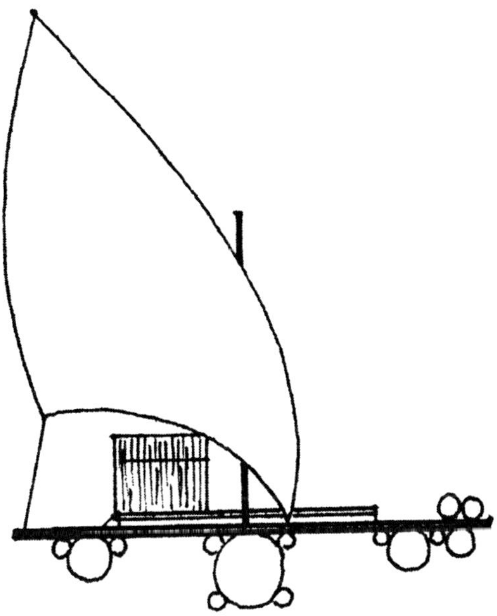

Fig. 7. De Bisshop's *Tahiti-Nui III* (after Danielsson 1960)

Under de Bisschops' guidance, the crew built a main framework balanced by a double-outrigger. The ten smaller ten-gallon tanks were divided equally to serve as the port and starboard outriggers. When this was accomplished, Brun built a box on the deck of *Tahiti-Nui II* that would serve as de Bisschop's sleeping area, and the explorer was moved into it on the afternoon of August 10th. The larger of the water tanks, the five forty gallon tanks, were then cut away from *Tahiti-Nui II* and pressed into service as the main hull of No. III.

On Monday, August 11th, 1958, Brun and his crew completed the deck, mast, steering oar, and additional crossbeams of the rescue raft. Only seventy miles from Starbuck Island, and after 180 days at sea on board the raft since its launch in Chile, the crew abandoned *Tahiti-Nui II* and set it adrift. The expedition had gone from its original conception as a voyage of science, to one of obvious pseudo-science, and now it had devolved to a struggle to survive.

But now it was too late to reach Starbuck. On August 16th, after fighting vainly against contrary winds, the *Tahiti-Nui III* passed twenty-nine miles south of the island and drifted on towards the central Pacific. They were now drifting too far north even to reach Tongareva, or Penrhyn as it was also called, and instead were looking at another thousand miles of open ocean between themselves and William Willis' landing place at Samoa.

When the raft missed Tongareva by forty miles on August 21st, another mutiny broke out, this time over food rations. The raft drifted along at about 9° south, and Brun knew that the Samoan islands lay between 13° and 16° south. De Bisschop, now semi-conscious, began to pray for death to take him. As Brun tightened the raft's lashings after a two-day storm, the other crewmembers listlessly ignored him. When the wind suddenly shifted around from southeast to northeast, Brun set course for the northern Cook islands of Manahiki and Rakahanga.

On August 30th, the reef at Rakahanga was only thirty miles away. Almost incredibly, Brun now proposed disassembling *Tahiti-Nui III* to build a smaller, fourth raft, to try to get the men safely over the coral. By four o'clock the next afternoon, the reef lay only ten miles off, and Brun was confronted with three choices: lay off the reef until they could attempt a landing in daylight; make for Manahiki, which de Bisschop believed was inhabited; or try to run ashore at Rakahanga at night. When the current, inexorably pulling the raft between the two islands, decided the issue, the men made ready to try to land over the reef at night.

They sighted Rakahanga around 5:15 in the afternoon. As darkness fell and the moon rose, the raft lay only a mile off the reef. As Brun vainly sought a passage between the coral, he knew they were engaged in the very "dangerous adventure [of] stranding on the weather side of a coral island" (ibid, 244). Just before nine p.m., the raft suddenly struck the reef, pitched into the surf, and the entire crew was thrown into the ocean. Alain Brun came to the surface and found de Bisschop bobbing in the surf, unresponsive. The other crewmembers managed to get the explorer ashore, and his third and final raft washed ashore as well around midnight. De Bisschop never regained consciousness. Despite the crew's attempts to revive him, his limbs began to stiffen around four in the morning. A medical examination would later reveal a severe trauma to the back of his head and a broken neck.

His last expedition and his life of adventure both at an end, the harrowing second leg of de Bisschop's radical experiment in drifting led to little serious discussion. Attention is most often focused instead on his failed bamboo drift on *Tahiti-Nui I.* Yet by setting himself adrift on a too-high trajectory into the Pacific from South America, de Bisschop had perhaps unwittingly showed the possible impact of accidental voyagers on an alien island.

Why, it has been asked, if they had truly reached the Polynesian islands in prehistoric times, did South Americans never transmit familiar cultural elements like stirrup-spouted ceramics or any other forms of pottery. De Bisschop showed that a raft crew, in extremis and attempting merely to survive, will lighten their load and increase their raft's buoyancy by throwing overboard anything they can.

Secondly, the culture-bearers amongst the voyagers—like de Bisschop with his vast knowledge of the sea and navigation—may have died en route. This would have the effect of leaving the survivors little in the way of a cognitive tool-kit with which to attempt to create a new and stable society. Even if the voyagers still possessed the knowledge to create pottery or voyaging boats, the raw materials for such works may have been unavailable on their newly-gained islands. Lastly, voyagers landing on a remote and sparsely populated island at the end of such a horrendous voyage would be lucky to be alive. They would have been unlikely, in such a state, to exert a profound effect upon a well-established population, other than to be quickly assimilated or eliminated. *Tahiti-Nui III* had washed ashore on an island with a population of about one hundred souls. The arrival created a great stir at the time, but it would be interesting to learn whether knowledge of it survives in any form today.

Searching along the shoreline in the days following the burial of de Bisschop, the natives discovered the metal jar in which Brun had stowed all the expedition's charts, films, and diaries prior to the wreck. All were intact. Soon after, a telegram arrived from Tahiti. The French authorities there had decreed that de Bisschop was to be exhumed, disinterred, and transported back to Papeete for burial. Taaroa would see him home after all.

Chapter 4

William Willis: Age and Adventure

Seven Little Sisters
Age Unlimited

Standing on the high platform, he looked over the expanse of low night fog above which, here and there, stood out the feathery heads of tall bamboo clumps and the round tops of single trees, resembling small islets emerging black and solid from a ghostly and impalpable sea. Upon the faintly luminous background of the eastern sky, the sombre line of the great forests bounded that smooth sea of white vapours with an appearance of a fantastic and unattainable shore. He looked without seeing anything—thinking of himself. Before his eyes the light of the rising sun burst above the forest with the suddenness of an explosion. He saw nothing. Then, after a time, he murmured with conviction—speaking half aloud to himself in the shock of the penetrating thought: "I am a lost man."

—Joseph Conrad
The Outcast of the Islands

Seven Little Sisters (1954)

No sooner had the account of Alain Bombard's raft voyage across the Atlantic appeared than another raft expedition took to the seas. In the summer of 1954, a lone American sailor, William Willis, tied himself to an incessantly rolling, pitching, sliding balsa wood raft in the middle of the Pacific Ocean. Unlike Heyerdahl or Bombard, who built or bought their rafts to test a hypothesis or two during deep ocean voyages, Willis had no scientific latitude to explore. His balsa raft followed no ancient prehistoric design. By his own admission, he had endured a lifetime of hard labor. Now, at sixty-one years of age, he sought one final test of his body and mind, as a way to prevent himself from being "cut off from that deep communion with Nature which meant happiness." (Willis 1955, 16)

A sailor and manual laborer all his life, he had gone to sea and around Cape Horn at the age of fifteen. He had been, in other words, a real deepwater four-master sailor before building his Pacific raft. "Raft-crazy" to him had always meant that bad thing that happens to sailors adrift for too long. But now he placed himself on a deliberate course to become raft-crazy. In a similar fashion as Bombard, he sought to explore just how far a human being could push their limits, and perhaps learn something new about survival at sea.

Willis wrote that the idea for a single-handed raft trip across an entire ocean came to him in 1951, as he toiled as an able seaman on a coal ship plying between Norfolk and New York. He had spent his life on such ships. As a young man, loading cargoes of nitrate along the western coast of South America, Willis had listened intently to the Chileans who told him of rafts that had drifted and sailed up the coast for centuries.

It is more likely that he had just closed the covers on *Kon-Tiki*, which had been published the year before. His career as a seaman had taken him into some of the same waters traversed by *Kon-Tiki*, and as a sailor he perhaps felt equipped to handle alone what the six non-sailors on board Heyerdahl's raft had accomplished. When he proposed the expedition to his wife, she used this very argument in reverse, pointing out that the six men on *Kon-Tiki* had always had their hands full in bad weather. How could he, at twice the age of the Norwegian adventurers, possibly believe he could handle such situations alone? Willis countered that only by going alone would he ever learn anything about the endurance of his body and mind.

When his wife agreed, Willis then convinced a manufacturer friend to support the expedition financially. By January of 1954, Willis was in the balsa tree forests of the Quevedo region of Ecuador, searching for balsa trees one hundred feet high and three feet thick. These were the same forests where Heyerdahl had harvested his logs in 1947. But instead of following *Kon-Tiki*'s replication of a pre-Incan raft, Willis, as a sailor, built a kind of modern sailing raft from these primitive materials.

> I would build a "hurricane" raft or no raft. I intended to follow no ancient design, evolved and proven by the masters of past centuries, much and profoundly as I admired them, for I had studied them and the way they were handled. I would build a raft that one man could sail and that would not come apart. Such a raft I knew had not yet been built. (ibid 1955, 38)

After innumerable false leads costing several months, and his window of time closing, Willis finally found a giant balsa tree near a hacienda along the Guayaquil-Quito highway. He found others clustered together in stands of three and four. They were draped with long, thin behuca vines which, he believed, "had been used since ancient times for tying the rafts together for the rivers and the sea" (ibid 1955, 41). When the woodcutters began taking down his trees in mid-March, sap flowed freely from the cuts. Willis ordered the workers to cut only machos, the male trees, which he believed were "much stronger though less buoyant than the lighter and more fragile hambres, or female trees" (ibid 1955, 46).

Willis then paid about twenty dollars apiece to have eight balsa logs—seven for his raft plus a spare in the event of an accident—cut and floated down the Palanque River to the estate of one Henri Kohn, whom Willis described as the "balsa king" of Ecuador. There he began to build his raft. When lashed together with one inch and a half manila rope, the seven logs formed a raft thirty-three feet long by twenty feet wide. Three hard mangrove beams were laid across the balsa, interspersed with smaller balsa beams. To these latter Willis attached a hut made of split bamboo.

To exert more control over the movement of the raft, Willis rigged up a kind of jib boom that projected from the bow of the vessel. This allowed him to use a jib sail forward of the raft's double mast. He considered the thirty-foot high double mast to be, like the identical mast Heyerdahl built for *Kon-Tiki*, "in the ancient pattern of all seagoing Ecuadorian and Peruvian rafts" (ibid 1955, 49). With several notable differences. First, Willis secured the double mast with steel cable. Then, by attaching its forestay to the protruding end of the jib boom, he was able to place the double mast much farther forward than any other raft. This allowed him the room to raise an aftermast that carried a large triangular sail. As a solo sailor, Willis needed to be able to control the movement of the raft from a single station. A man alone could never manage the big sweeps, or steering oars, that Heyerdahl and his five companions had employed on *Kon-Tiki*. So in this instance he forsook ancient technology in favor of "a regular ship's wheel connected with a regular rudder" (ibid 1955, 52).

Unlike the estimable Dr. Bombard, Willis had no desire to starve himself deliberately. He would take plenty of rations, but these would be his version of the minimum amount of food a man needed on such a voyage, supplemented by whatever fish he might catch. He had made no calculations, as had Bombard, of the composition of fish, or how many pounds of fish flesh one would need to squeeze in order to extract live-giving fresh water. Instead,

Willis devised a diet that consisted in the main of foods he that considered had sustained the prehistoric peoples of Peru.

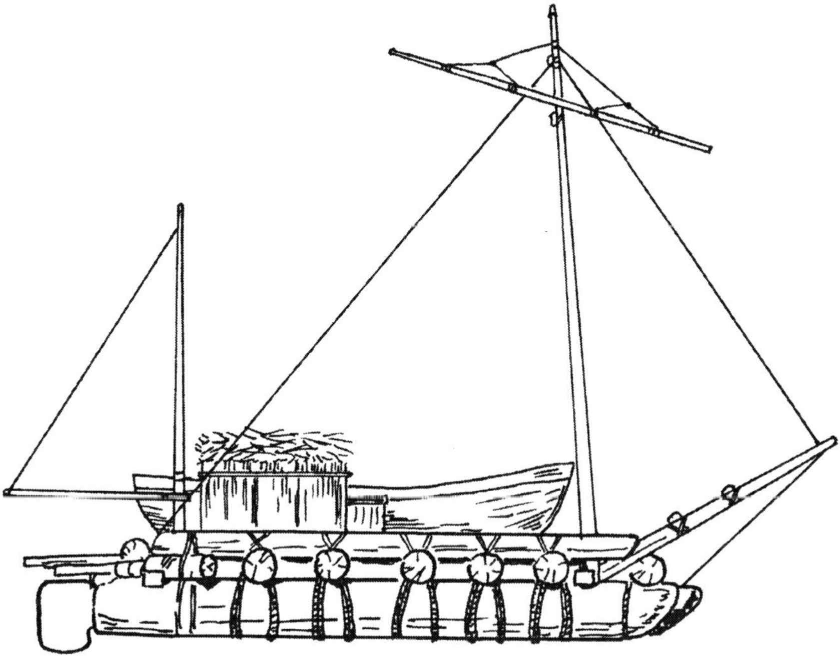

Fig. 8. Willis' *Seven Little Sisters* (after Willis 1955)

He had prepared for him fifty pounds of a flour made from roasted and ground cañihua, a cereal that was found in the high Andes. Along with seventy pounds of another flour roasted and ground from barley, which the natives called machica, and another seventy pounds of unprocessed molasses-filled raw sugar called raspadura, Willis had his prehistoric provisions. He would prepare these staples by mixing them with a little water, kneading the result into a ball, and eating it. As increasingly he came to rely on this mixture during the expedition, he arrived at a belief that a diet "of nothing but whole grain ground into flour was the secret of survival in Europe and Asia throughout the ages, during thousands of famines that swept over the lands in the wake of wars or bad harvests." (ibid 1955, 83)

As an emergency raft, Willis obtained a twenty-six foot dugout canoe made from red cedar. For communications, Willis would carry a radio and a generator to power it. When the British owner of the Marconi Company, Sir George Nelson, on a visit to South America, heard that Willis had not the time nor funds to ship an emergency transmitter, Nelson saw to it that one was provided for free. Yet these concessions to the modern world were of little relevance for Willis, because he saw his goal as something higher than mere anthropological or scientific experiment.

> …I did not want to prove any scientific theory, or discover and set up any new course of any kind for others to follow. To me, this voyage was something much more—it was a pilgrimage to the shrine of my philosophy. Call it an adventure of the spirit. On this voyage I wanted to prove—had to prove to myself—that I had followed the right star throughout my life (ibid, 15).

Willis was so confident in his new type of ancient-modern polyglot raft, he thought he could head directly into the Pacific from the port of Guayaquil in Ecuador. From there he would catch the Humboldt Current that sweeps from the coast of western South America across the eastern Pacific to Polynesia. But both local fishermen and a small knot of European and American adventurers, beachcombers, and treasure hunters all warned him off this route. It was by now winter in the southern hemisphere, and the Humboldt Current, rather than bend toward the Pacific as it careened off Cabo Blanco 200 miles south, would instead flow north before turning west. Willis would be driven towards Columbia or, at best, the Galapagos islands 600 miles off the coast of Ecuador.

Only the year before, Thor Heyerdahl's archaeological expedition to the Galapagos had discovered potsherds in the islands later identified as pre-Incan in origin. For the first time, concrete evidence had been unearthed that prehistoric mariners had made the trip from South America to the nearest offshore islands of the eastern Pacific.

One night in a local restaurant, Willis met a Swiss archaeologist who warned him off his intended point of departure as well. Once at sea, he felt that the currents would drive Willis's raft north along the coast, to the treacherous Bay of Panama instead of Polynesia, along what many presumed was the prehistoric raft route to Central America. Willis countered that maybe he could get towed to Cabo Blanco, and leave from there. But the archaeologist said it was too late for that. *Kon-Tiki* had departed from Callao in Peru, 700 miles south, in early April. It was now June. Willis was running out of time.

"The Indians," the archaeologist went on, "went down to Paita, a little port about halfway between Guayaquil and Callao. That's where they started from to get to the Galapagos. And they knew just when and how to sail. They knew the seasons and the moon and the wind, things we don't know anymore. And they had crews to handle the rafts. Nobody went alone" (ibid 1955, 55).

Willis was told to wait until next year, but the thought of seeing his raft disintegrate into the mud of a South American river was too much. He managed to get the raft towed to Punar, where a Grace Lines ship would take her on board for the voyage to Callao. At Punar, however, the raft was almost rammed by a gasoline barge. As he fended off the barge from crushing his raft, Willis suffered a hernia. He told no one, for fear that they would insist he be hospitalized and his Pacific expedition stopped.

In Callao the same Peruvian naval base where *Kon-Tiki* had been built seven years earlier was put at Willis's disposal. In Callao, Willis stowed his provisions, stepped the double mast and rigged the ship, fastened the sails, and brought on board the radio and transmitter. He rejected all sailing volunteers in favor of a cat and a parrot. After a brief test run of a few hours, Willis proclaimed the raft ready for sea. In honor of the balsa logs he prayed would deliver him across the Pacific, he christened the raft *Seven Little Sisters*. In fact, the raft performed so well that his wife made him swear he would indeed stop at his planned destination of American Samoa, and not attempt to cross the entire Pacific to Australia.

Like *Kon-Tiki* and *L'Hérétique* before, the raft *Seven Little Sisters* could not make its own way to sea from its berth at the docks. On the 22nd of June, 1954, the Peruvian tug San Martin towed the raft sixty miles offshore, west of Callao, and let go the tow rope. Willis' last visitors were three native Peruvian sailors, who looked dubiously at the raft. Willis asked them if they thought he could make it.

> I looked from [one] to the other two seamen, obviously descendants of men who had put out during past centuries from these same shores and in similar rafts, who had sailed from Peru up the coast to Panama and to the Galapagos, and in some instances had been driven helplessly by storms or, to avoid extermination by enemies ashore, had sailed westward into space, taking the same trail that lay before me now. (ibid 1955, 70)

Willis had insisted that the Peruvians clear him for America Samoa, like any regular ship. After some trepidation, the authorities gave in and he got his wish. Now all that separated him from his goal was seven thousand miles of open ocean.

Seven Little Sisters entered the Humboldt Current and began to drift northwestward away from Peru, in late June, 1954. As the balsa wood raft lolled through mild days and cold nights, Willis felt himself drifting as well—away from one reality and into another. "How strangely everything drifted out of my mind," he wrote, "as if I had had no real contact with the world of men, no real hold on its existence. I could hardly remember anything and details not at all" (Willis 1955, 79).

His sense of time disappeared. For the most part Willis ignored his radio and transmitter, wanting only the elements of the ocean and the sky to know where he was. He imagined losing his rigging in a storm and being forced adrift for months, even years, all the way across the Pacific, alone with his cat and his parrot, while his wife and friends mourned his loss. As time began to seem more and more insignificant, Willis felt that he had seemingly rafted into a dimension of experience where time could not enter. He felt disturbed, as if a great big sea had swallowed him and his raft, and he began to doubt his ability to navigate a way home.

William Willis built his first raft at the age of four. It is true that I admire most those explorers who set themselves the task of testing new ocean-going hypotheses on old problems of human exploration and migration. But I also have immense depths of sympathy for those, like Willis, who simply wanted to get lost in time, and sense great natural truths around them, as they rafted on archaeological seas. That Willis attempted his solo expedition across the Pacific at the age of sixty-one, when most life trajectories are setting toward the shores of retirement, is all the more remarkable.

Perhaps it was a lifetime of sailing that enabled Willis so quickly to enter a kind of spirit world as he rafted offshore from

Peru. Unlike Heyerdahl, he already knew how the set of his mainsail or aftersail would react in various winds; unlike Bombard, he did not need to invest the first weeks at sea in teaching himself to navigate. Finding a roll of white medical tape in his kit, he wound it around the 'midship spoke of his raft's steering wheel. As he explained, the 'midship spoke would appear at the top of the wheel when the rudder it controlled was parallel to the raft's center log, exactly as it would show on a regular sailing ship when it was parallel to the keel. No matter where he was on the raft, he would be able to look at the wheel and know at what angle the rudder was taking the seas.

Alone now, he knew he was on trial, and that fact only intensified the determination of this most interesting of all American solo sailors. "I leaned over the side and dipped my hand into the sea and sprinkled the water over my head and face and chest. I had done it on impulse, not thinking what I was doing, and then I realized that it was a sort of baptism, a consecration" (ibid 1955, 73). He found, perhaps metaphysically as well as in reality, that there was no straight course for a man alone on a raft. The raft swung from side to side, and would 'run-off' on stubborn courses of its own. It had to be attended constantly, for if Willis left the helm for only a moment, the raft would heel over and set off on its own. Still, he wished that the ancients were somehow watching him make his way across the Pacific on one of their balsa rafts. He imagined a Pacific where a thousand ancient rafts had preceded his.

Ten days out from Callao, neither of the two stoves on board would light, the radio wouldn't work, and the chronometer died. Willis took to mixing his flour and eating it uncooked. He made his instant coffee with cold water. The wind was steady and always cool. A shark took to following the raft, and Willis was at first uneasy about its presence. The shark trailed just behind the steering wheel, and to Willis' imagination it was only waiting for the old man to slip and fall overboard. One night Willis tossed a fish overboard just to see if the shark was still around. Out of the blackness shot the fin and tail, seizing the fish and disappearing back into the depths.

Every day became a new lesson in sailing his polyglot raft. He knew that hurricane season was approaching rapidly, so he obsessed upon getting into the Pacific farther and faster than had Heyerdahl and *Kon-Tiki.* He knew that if strong winds swept his rigging overboard, he would be sentenced to an uncontrolled drift voyage that might last years.

And he continued to withdraw into his own imagination, with the spatial replacing the temporal. The faces of old friends began to dim, to be replaced by the nightly visits of the planet Venus and the constellation Southern Cross. Even things happening now on the raft held little importance to him. It seemed that he was "lost in time, time which had come to mean nothing. Days or months, what were they?...I was in a void where time could not enter…I sensed immense truths around me in space, revealing themselves in beauty and yet shrouded in eternal mystery" (ibid 1955, 97; 119).

Nearly a month out from Callao, on the 17th of July, 1954, a mysterious and intense pain formed in his solar plexus. For several hours, he felt the whole front of his body tightening around this knot of pain. He took aspirin, bicarbonate, even brandy, but nothing stemmed the agony. He had told the authorities in Peru that he would try to sail between the tenth and twelfth parallels of latitude during this stage of his voyage, "following the general course taken by the *Kon-Tiki* expedition some seven years earlier." (ibid 1955, 123) But as he suffered through his pain, he found he was much farther north, only three degrees south of the equator. He was far out of the shipping lanes, and he doubted that an S.O.S. sent from his small transmitter would be heard in Peru. Even if it was, who could rescue him now? As the pain subsided temporarily that night, Willis managed to record some of his most memorable thoughts.

> All the stars were out. The Southern Cross, tipped forward, stood blazing on my left, like a beacon hewn out of space. A little behind it and above rose the big southern constellations amid myriads of lesser lights that looked like cities and highways on the slopes of eternities. (ibid 1955, 122)

The pain intensified the next morning. Willis made a last meal for the cat and the parrot, wrote a last message to his wife, and crawled into his cabin to die. He scanned his past, searching for some sin he must have committed in order to justify this infliction. He thought maybe he had wrongly challenged the gods by attempting the impossible, and was now being revealed for the mortal he was. When at last the pain became unbearable, Willis feebly cranked out an S.O.S. message on his transmitter. He had decided to give up. After that, he crawled to his cabin, made a mugful of aspirin paste, and choked it down. (Only after his voyage was over would Willis learn from a doctor that he was suffering from a perforated stomach ulcer.)

He lay down in his cabin to try to sleep and suddenly felt the knot of pain begin to lessen. Poking around his ribs, he found that it was true, the pain had fled as mysteriously as it had come. Hurriedly he scrambled back to his emergency transmitter, and tapped out a new message: "All well … Need no help … Need no help … Need no help." Whenever he felt strong enough, he continued to tap out the same message. The raft, with Willis suddenly renewed, continued its voyage into the Pacific.

Willis celebrated his rebirth by beginning to believe that he had been spared death for a reason. He felt that if he were allowed to drift on and on he might finally reach a kind of understanding of all things. Barring that, he could at least continue on to show himself that he had lived a good life, had not squandered the gifts bestowed upon him. He had a sudden and blinding premonition that, in this young nuclear age, all humans had somehow perished and he alone had survived, to wander forever between sea and starlight. Perhaps he was condemned to drift from island to depopulated island, the last living soul on earth.

Willis began to consider more mundane eventualities as well. Now that he was approaching the halfway point of his voyage to Samoa, he began to realize that it would be nearly impossible for a single individual to land a ten-ton raft on a reef-encircled island. If he could find some way to moor the raft temporarily, he could use his dugout canoe to scout through channels in the Polynesian reefs, and perhaps find locals who would return with him to the raft and help guide it to a safe anchorage.

No sooner had Willis thought this problem through, than he discovered that the seams of nearly all his five-gallon fresh water tins, stored beneath the deck of the raft, had been corroded by salt water. With the exception of three tins, his entire fresh water supply was gone. Even the three intact tins began to leak, and Willis had to frantically scramble to salvage about nine gallons of fresh water.

The possibility of dying from thirst now became very real, and so Willis resorted to an old trick he had learned as a boy on board tall ships. To stretch his fresh water supply, he began to drink one or two cups of seawater each day. He had seen the value of it himself when an old cook on a sailing ship had been cured of constipation by drinking a regular cup of seawater. In this he seemed to know already what Alain Bombard felt necessary to learn by crossing the Atlantic on a rubber boat. Willis felt he could drink salt water fairly easily because he was a vegetarian, and therefore had no real craving for salt. So the seawater doubled as a daily salt ration.

The raft plowed westward at sixty, seventy, eighty, sometimes ninety miles per day. On calm hot days, Willis would sit near the double mast, dip his feet into the limitless blue water, and feel the wind as it careened down the big square sail. Whales often visited the raft, approaching its wake from the southeast and swimming alongside it for a time, before making off for the northwest. By now, in mid-August, as Willis neared the Marquesas islands, he prepared to steer well clear of the archipelago's high volcanic cliffs. To hit the shelterless Marquesas he considered suicide, though he briefly toyed with the idea of attempting a landing in order to replenish his fresh water supply.

When the sea snapped several of the raft's centerboards, Willis was forced to take up the catwalk leading to the bowsprit, as well as several of his deck planks, and make new centerboards from them. A green forest of mossy marine growth had appeared on the underside of all the balsa logs. On September 1st, 1954, Willis recorded his position as being directly north of the Marquesas. Birds flew out to alight on the raft during the day, only to return to the islands as night. The same shark still followed the raft, still a constant companion after 2,500 miles.

The general calm of the passage to this point was interrupted by a sudden and fierce storm, one that tore away the mainsail and left the raft powered by only the jib and aft sail. The raft dug so deeply into the troughs of the waves that Willis thought he would have to lash himself to the deck to keep from being swept overboard. When the storm blew itself out, Willis noted that the balsa logs, after four months of constant immersion, were now riding four inches lower in the water than when he started. Yet he felt fortunate. If he had equipped his raft with one sail, as he felt the ancient rafts of Ecuador and Peru had been rigged, he would now be dead in the water. Instead, the raft was still making a good heading under the fore and aft sails he had insisted upon.

Willis felt that his vision in designing the raft was vindicated when, on September 5th, he studied his charts and found that he was directly north of Raroia. His raft had passed the same longitude as the *Kon-Tiki* had achieved when it concluded its voyage on the tiny island in the Tuamotus in 1947. Moreover, the *Seven Little Sisters* had accomplished in seventy-four days the same 4,300 miles it had taken *Kon-Tiki* 101 days to cover. With justifiable pride, Willis now entered what he considered the "home stretch," the long downhill drift from the Marquesas to Samoa. He felt now more than ever as if he existed "in a prehuman world, where man had not yet been created—alone in a world of water and stars and suns and roaming winds" (ibid 1955, 194).

Around ten o'clock in the morning on September 18, 1954, Willis sighted uninhabited Flint Island. He climbed to the top of the double mast to see a strip of white sand, covered with a solid line of trees. After 5,500 miles at sea, it was like a vision, and seemed to call to him to abandon his pilgrimage. Willis sailed closer and closer, until he could hear the sound of the surf breaking on shore. But he saw no simple place to land, and feared that a current setting toward shore might grab hold of the raft and pull him in, and he would be marooned. Reluctantly, Willis swung the raft off, and Flint Island, "lonely as the mass grave of a forgotten race" (ibid 1955, 215), fell off behind him. Ahead, a thousand miles away to the southwest, lay Samoa and the way home.

Less than a month later, after 112 days adrift from Callao, Willis sighted Tau, the easternmost Samoan island. His attempted landing—on Columbus Day, of all days!—was thwarted by impassable reefs. So Willis made for Tutuila, the main island of American Samoa, and location of the main city of Pago Pago. Once again, currents and a lack of landing places amid the reefs forced him westward. It was not until a local American vessel tracked the raft down and put it under tow was Willis brought safely ashore after a voyage of more than 6,700 miles.

In Samoa, Willis learned that he had been considered lost at sea for almost four months. His appearance off Tau, combined with the brief radio message he had sent to the authorities, were the first signs of him since he had been cast adrift off Callao in June. Willis had made one of the greatest solo sea voyages ever recorded.

After the American governor welcomed Willis to the islands, and introduced him to the local Samoan chiefs, Willis promptly donated his raft to the people of American Samoa as a permanent museum exhibit. After being jostled ashore, the raft was reassembled in a prominent place of honor along the harbor at Pago Pago. The government struck up a bronze plaque celebrating Willis' historic—and perhaps prehistoric—voyage upon it.

Barely ten years later, when Willis returned to check on it, the raft had vanished. One person said it had been chopped up for firewood, but no one could recall just what had happened to it.

Age Unlimited (1963)

> I believe that he who has been in lonely seas will go back for the peace he knows he will find. But it is not his for the mere asking. Each day, each moment, must by the nature of his being be one of suffering, for he has to tear out what binds him to his kind and above all to his father and mother, his wife or child—the blood anchors which hold him to time and space—and stand naked and shorn in anguish. At last, if it is so willed, he may gaze into the solemnity of silence and see himself. If he becomes frightened at his own smallness, he screams for help until he becomes mad. Then the revelation is not for him.
>
> —William Willis
> *Whom the Sea Has Taken* (1966)

William Willis returned home from his 1954 balsa raft expedition and almost immediately began to feel the after-effects. Whether from age—he was now sixty-four—or from the severely restricted diet and exertions of the Pacific expedition, he found himself suffering from diabetes, arthritis, stomach and kidney problems, and a paralyzing nervousness. All his life he had been a kind of health freak, and more than that, someone who deliberately sought to push his body to new and greater extremes. Now he felt his body begin to collapse away, and the shock of that collapse frightened him into action.

He now called upon a lifetime of hard physical labor, from a spartan boyhood along the Elbe River in Germany, to backbreaking exertions as a seaman loading cargoes onto sailing ships around the world. All his life he had sought not only physical strength and speed, but also the mental toughness to withstand pain, hunger, and thirst. He had endured a hard boyhood, first from fatherly abuse and later from the deprivations caused by the final days of merchant sailing, as greedy owners cut corners, wages, and rations, to preserve their shrinking profits. Willis had always admired the Indians of North America for their ability to endure painful initiation rites, and had more than once been forced to draw upon this admiration for his own survival. Now, at an age when most men are either retired or dead, William Willis sought instead to redouble the strain upon a weary body and mind.

After lecturing about his raft expedition in the U.S. and Europe, Willis moved with his wife to the hills of southern California, ten miles from the Pacific. There he subjected himself to years of natural healing therapies. In this, he anticipated by more than a decade the natural foods and holistic health movements. He fasted and meditated, then went on diets restricted to food grown in his small garden plot. "I did breathing exercises, I took water cures and other treatments, and rigged up a steam bath; I drank herbal teas and sometimes only distilled water; for weeks I lived only on cereals, then only on fruits and vegetables" (Willis 1966, 34-35).

At last he felt he had recovered the vitality his body seemed to have lost. Yet there was something more that he required, something that his spirit was now lacking. As he sat on the beach one morning, looking out across the Pacific, he began to realize how much the 115 days he had spent alone on his balsa raft *Seven Little Sisters* had affected him. He found himself longing for "those glorious days and nights when the whole world was mine, the sea and the sky from the highest star to the dark depths beneath, with not a soul near of all the earth's millions" (ibid, 35). He recalled the exertions of carving a wooden raft out of the jungles of Ecuador, of drifting on calm seas with sails hanging from idle masts and logs growing foul with moss. He remembered the slow rusting of his water containers, and how a torrential rainstorm north of the Marquesas—the only rain of the voyage—had saved his life. The remembrance of these transforming events in his life made him realize suddenly that he had to make one more raft voyage.

"I knew … that I would want to leave from Callao as in 1954, and I knew that this time I would want to sail nonstop to Australia, to Sydney if possible. It meant crossing the whole Pacific, from east to west—at least eleven thousand miles of sailing, alone again and without touching land. Was it a dream begotten in silly idleness never to become reality? I asked myself a hundred times. Could it be done?" (ibid, 39).

He could easily have asked, Could it be done by a seventy-three-year-old man alone? For Willis had no intention of returning to a raft if it meant taking along a crew. He was convinced that taking anyone with him would have "watered down the expedition to a routine job, even to drudgery" (ibid, 72). By now he had almost certainly read Eric de Bisschop's account of the *Tahiti-Nui I* expedition, and Bengt Danielsson's transcription of Alain Brun's account of the *Tahiti-Nui II* disaster. The thought of trying, like de Bisschop, to meld his unique personality with that of a man one-half or even one-third his age, on a voyage lasting months, was not one that Willis would have considered for very long.

Remembering his arduous search for suitable balsa trees in the interior forests of Ecuador, Willis concluded that, even if he could drift across the Pacific, there was little hope at his age of duplicating such a jungle expedition. Willis thought that lengths of pipe, welded together to form a unique kind of trimaran, would make appropriate substitutes for balsa logs. He returned east to New York, to visit with many of the companies that had helped on his first expedition. So it happened that the raft for this greatest of Pacific solo raft adventures was built in a boatyard along the Passaic River in New Jersey.

"The raft consisted of three [metal] pontoons [filled with polyurethane], each twenty feet long—two aft and one in the center and each having a somewhat flat shape like the hull of a power boat. The length at the waterline was approximately thirty-four feet and the width about twenty. A six-inch heavy steel pipe encircled the three pontoons on top and, together with girders running from side to side, welded the whole into a solid mass" (ibid, 43).

Over a pine deck Willis built a seven-foot-long steel cabin for his navigating instruments and provisions. The mainmast was thirty-eight feet high and hauled an eighteen-foot wide yard for the mainsail. By early May, 1963, the whole polyglot contraption was ready to be towed down the Passaic River to Newark. Moving at a stately pace on the still creek, Willis was dismayed to find the raft top-heavy, the deck awash. He would have to cut away the steel cabin and replace it with bamboo once he was in South America. At the Newark docks, Willis showed off his new mainsail, painted with the name of his new raft: *Age Unlimited.* Then the raft was lifted on board a Grace Line steamer bound for Callao.

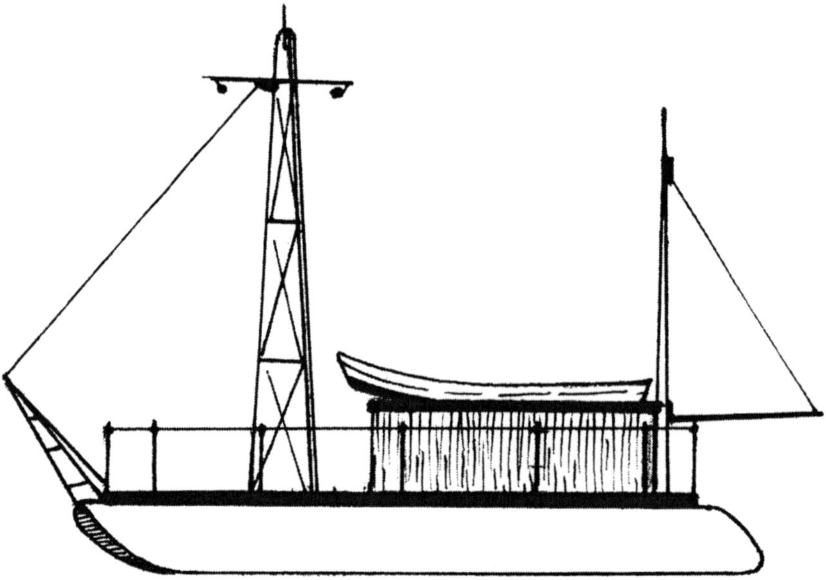

Fig. 9. Willis' *Age Unlimited* (after Willis 1966)

On the trip to Peru, Willis brushed up on his celestial navigation. He had not taken a sun sight since the 1954 expedition, and had never navigated east of Greenwich, on the other side of the International Date Line. Yet if his raft voyage succeeded, he would cross that line—halfway around the world from England—somewhere near Fiji. From that point on to Australia he would be drifting through waters completely unfamiliar to him.

In Callao, Willis sorted out affairs both political and cognitive. He was visited by a man carrying a bulging scrapbook of articles cut from local newspapers about the 1954 voyage. The man also possessed a similar portfolio from the raft expeditions of Eric de Bisschop. Willis wrote a letter of permission for this "dreamer of raft voyages" (ibid, 56) to visit his raft, since the military junta ruling Peru had denied any local access to Willis or his vessel. Willis then sat his wife down and convinced her that they could communicate telepathically while the raft drifted thousands of miles at sea ("distance means nothing in electricity and in thought even less" (ibid, 59).

On July 4, 1963—an exceedingly late departure date—the Peruvian navy tug Rios towed *Age Unlimited* to sea. Several miles offshore, Willis suddenly felt the raft slide into an immense trough, as if it was falling off the earth. The Rios was nearly overturned. The towline did not part, however. When the ocean had stabilized somewhat he was told that a submarine earthquake had shaken Lima and Callao to their foundations, and caused a tsunami that had lifted and then dropped both the raft and the tug on waves several hundred feet high. It was a terrifying beginning to the expedition.

When the towline from the Rios was finally let go and Willis drifted away, he received his first impressions of how the raft would behave, and he was not pleased. *Age Unlimited* rolled and jolted itself with every wave, pitching forward, then rolling side to side. "Its motion was violent and continuous and quite different from the action of a boat, and therefore it put a great strain on me to retain my balance…

"A well-built and properly rigged boat often holds its course for one, two, and even three thousand miles without a hand touching the wheel or tiller… On a square-rigged raft one has the sheet, tack, and the braces to handle, and the steering has to be done standing, since it requires considerable strength which cannot possibly be applied while sitting" (ibid, 70-71).

Willis knew perfectly well that a large raft, rigged with main and aft sails as his was, required a three- or four-person crew to handle. But he had done it alone once before, and he was determined, despite the constant pitching and rolling, to do it again. He would bring back as well a testament to human endurance, an inspiration as much for out-of-shape youth as for senior citizens.

Five hundred miles from Callao, Willis encountered his first setback. Both of the tillers that controlled his rudders were cracked. He knew he had to get to shore to repair this damage, and the only logical choice was the port of Guayaquil in Ecuador. Willis even toyed with the idea of scuttling his rolling metal trimaran in favor of a new and stable balsa raft, especially when his new creation rolled "from beam to beam at its regular rate, smashing its sides into the sea as if it wanted to pitch the masts out of their sockets and overboard" (ibid, 74). But by July 16th, a landing at Guayaquil became out of the question. The raft was stuck in the Humboldt Current, and could not make it east toward the coast of South America. Willis could only hope that the rudders would hold together long enough for him to clear the Galapagos and make for a landing somewhere in the Marquesas.

Feeling "a bit stuffy around the middle and sluggish mentally" (ibid, 80), Willis began to consume about a cup a day of seawater. Like his fasting and health food diets, he believed that a daily dose of salt water promoted both mental and intestinal regularity. His raft was another matter. Each time he tried to jury-rig a solution to his rudder problems, heavy seas smashed all his hard work. He did not want to be reduced to steering with only his sails and centerboards, fearing that this would ultimately be fatal when navigating amongst the islands and atolls that separated Samoa from Australia.

As the weeks rolled along, Willis imagined the faces of his friends fading from his memory. By mid-August, he was seeing his wife appear as an apparition on board the raft. Then his long-deceased mother began appearing to him as well. Again, as on his earlier voyage, the mental trials seemed infinitely greater than the physical. When he grew tired of the performance of his mainsail, which through a bollix in the design hung more than eight feet above the deck, Willis simply stitched a new sail and rigged it underneath the main. He even started to sketch out a new raft, with even more sails, which he thought he could build for future expeditions. Even more prescient, Willis imagined his raft as "an ideal place to make experiments" (ibid, 111). It was a notion that would occur to at least one social scientist in the 1970s, who set out with four other men and six women to test human compatibility on a raft voyage across the Atlantic.

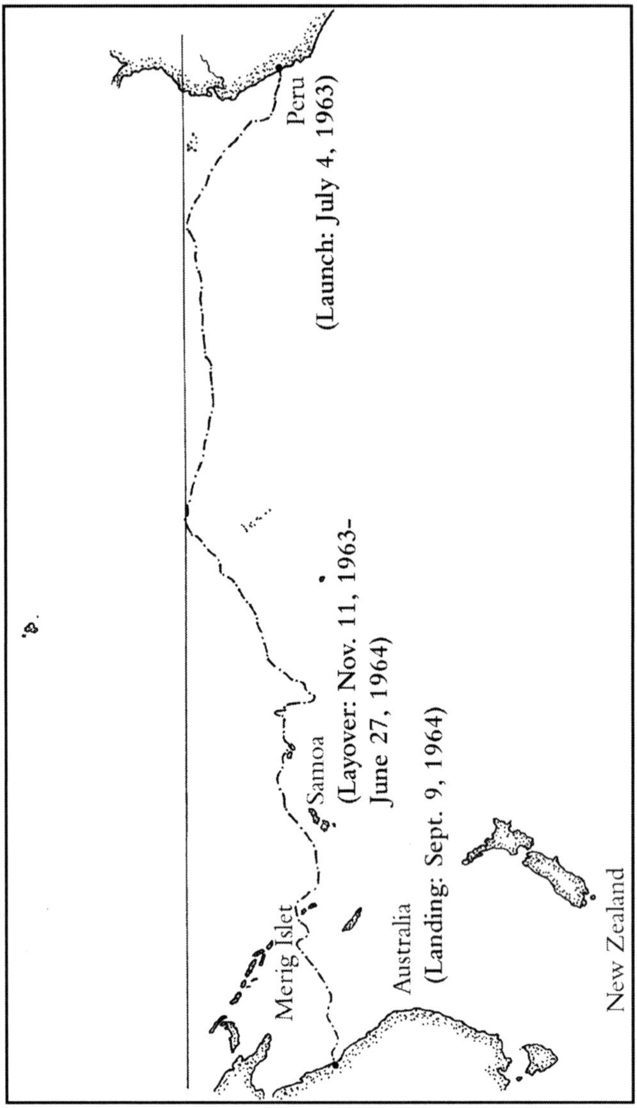

Fig. 10. Route of the *Age Unlimited* from Peru to Samoa, and Samoa to Australia (after Willis 1966)

In early September, Peru lay 4,000 miles behind, and Pitcairn Island, refuge for the *Bounty* mutineers, 1,400 miles to the south. Willis was tormented for several weeks by a premonition that his wife had been killed in an automobile accident. On September 11, 1963, the raft passed more than 300 miles north of Nuku Hiva island in the Marquesas. Comparing this position with his 1954 raft expedition, Willis found himself 140 miles further north than his earlier drift. He attributed this to the damaged rudders, which he had not been able to fix properly. *Age Unlimited* was now just three degrees south of the Equator, and Willis knew that Filippo Reef, which spread from five to six degrees south, was only about 700 miles away to the west. He had to get the raft turned toward the south or the expedition would be wrecked on the fifty-mile-long reef.

Two weeks later, about 700 miles north of Tahiti, Willis climbed to the top of his mast to scan the horizon for the breakers of Filippo Reef. Just as he prepared to hit the reef, the wind shifted around to the northeast and pushed the raft to south of the breakers. A voice followed by an apparition began to appear to him daily, telling Willis he was doomed and that his only hope was to jump overboard. He thought his restricted diet ("diet is life—man is what he eats" (ibid, 129)) might be getting the better of him, forcing nightmares upon his lonely mind. He dreaded becoming mentally incapacitated. Seventy years were, after all, seventy years, and he could not be expected to last forever. How much longer did he really have? Ten, twenty, thirty years? It was really not much time in the stream of eternity, and he was acutely aware of it.

Such thoughts of death came with even greater frequency as the raft arced just south of the 12th parallel to be sure of clearing the reef at Rakahanga where de Bisschop's neck was snapped in the surf. The weather was continuously worse than the mild summer of 1954. Willis cut open a shark to get at the vitamins its liver possessed. When he went over the side to make another attempt to fix the rudders, he tossed out the shark carcass on a long line, to keep the living sharks away from his legs. With his fingers trying to blindly splice cables as his nose stuck out of the water for air, Willis felt something give in his belly. He managed to crawl back on deck, only to discover that he had ruptured himself of the left side, to match the right hernia he had suffered on the 1954 expedition.

By late October, as he began to fear an encounter with former cannibals of the New Hebrides, he realized that Australia was fading further and further from his grasp. Between his rudders and the new hernia, he had to get ashore somewhere and repair both. One hundred and thirty days out from Callao, he spotted the mountains of Upolu in Samoa, where prehistoric mariners had settled 3,000 years earlier, and where Jakob Roggeveen, the first European to see Easter Island, had landed in 1722. In 1962, a year before Willis' raft came in over the reef, Samoa had become the first independent island nation in the South Pacific.

Pounding over the reef into a protected lagoon, the raft was surrounded by rocks the size of slumbering elephants. Willis had come some 7,500 miles in 130 days, farther than any solo raft expedition before or since. He was met by a Methodist missionary in a small skiff, who seemed frightened at the specter of the white-bearded, half-naked old man on the raft. Willis beckoned to him that he was fine, but could the man please send word to his wife in New York that he was okay? When the missionary left, Willis sat back down and opened a cold can of beans. He had accomplished one of the great drift expeditions ever attempted, yet he "felt suddenly lonely, as if something has been taken out of my life" (ibid, 162). He knew he would never be able to rest if he did not finish the rest of his voyage to Australia.

As he assessed the condition of both his raft and himself after the landing in Samoa, William Willis knew that both required substantial overhauls. The rudders of the *Age Unlimited* had to be reconfigured for the long southward drift towards Australia, and Willis himself had to get to a doctor. He was diagnosed with a hernia, and the doctor warned him it could strangulate and kill him during his proposed voyage to Sydney. He cabled his wife (who had not died in an accident after all) and she insisted he come back to New York for a check-up there.

Willis gave his raft to the government of Samoa, in the event that he was not able to return and take up his quest. Always thinking ahead, he feared he might be killed in a plane crash on any of the many flights to and from New York. Or on the operating table if his doctors forced him into surgery. In fact, Willis' New York doctor gave him the same diagnosis and prognosis as the doctor in Samoa, yet Willis resisted all medical advice. He feared that to stop for a major surgical procedure now would mean the effective end of his expedition: as he slowly recovered from an incision to his belly, his metal raft would rapidly rust in its Samoan lagoon. As he told his wife "I wouldn't be ready for months, and by that time the pontoons will have rusted through and the raft be lying on the bottom of the lagoon" (Willis 1966, 174). By the time he returned, if he ever did, the raft would be too far gone to salvage the completion of his journey.

Rejecting his wife's plea to take her on the remainder of the drift to Australia, Willis returned to Samoa to find the raft rusty and his water barrels fouled. He took up residence in a small bungalow, and got to work scraping and sanding, and tending to the repair of the rudders. He fixed his departure date for June 27, 1964, but Willis was confronted by a host of last-minute problems. Cockroaches had infested his wooden water barrels, and despite numerous attempts to clear the containers of bacteria, eggs, feelers, and other sundry insect parts from his water supply, nothing seemed to work. He finally settled on a handy supply of chloride with which to treat the water if it became too foul on the voyage.

A cockroach of a different sort hounded him on land. An itinerant Pacific schemer with a young Hawai'ian wife—the latter obviously made an impression on Willis ("as slim as a young seal with black dagger sharp eyes who wore incredibly tight Chinese dresses" (ibid, 185)—planned to follow and film the raft, and sell the results in the U.S. If successful, the poaching would kill Willis' own documentary, along with the money his film would raise to cover the cost of the expedition.

The threat was enough to force Willis to sea at night. A local tug was recruited to tow *Age Unlimited* thirty-five miles offshore under cover of darkness, and let go the line. At nine o'clock in the morning, June 27, Willis was again at sea. But his voyage soon turned to near-disaster. The raft had not been towed far enough offshore, and began to drift back to the reef at Samoa. Worse, as he tried to signal an S.O.S., a stinging pain on the left side of his abdomen brought all the worst fears of his doctors to the surface. The exertions and stress of getting the raft ready for sea had only exacerbated the rupture he hád suffered on the voyage from Peru.

> I tried to get the hernia back in, meaning the pinched-off part of the intestine, while lying on my back with my legs up to relax the abdomen. I tried every form of manipulation, at first gently, then, as I became desperate, more forcefully until I realized it might aggravate matters. My whole abdomen had become distended. Now and then I looked at the mountains of Savai'i lying in the glassy sea beside me, but they had lost their terror, since I had a greater danger to contend with. (ibid, 195)

As the raft drifted closer to the reef, Willis was forced finally to running a rope through a block, tying one end around his ankles, and stringing himself up to relieve the pressure on his abdominal muscles. After several hours of twisting and turning in an attempt to loosen the obstruction, the pain at last gave way, he felt the protrusion retreat, and a sudden wind from the east pushed the raft clear of Samoa and at last on its way. Willis rigged up six different sails to catch all possible wind, and started on his downhill run to Australia.

The pain from his hernia had Willis trussing himself upside-down for hours a day, like a side of beef with a shrinking twenty-six inch waistline. He watched a coconut bob past the raft, carrying two agile crabs as cargo. It caused Willis to wonder about the migration of animals and plants around the Pacific. In mid-July, Willis rounded Fiji and tried to force the raft south of New Caledonia. But persistent winds from the south were pushing the raft north and west, away from his intended route southwest toward Sydney. If he couldn't get the raft pointed south, he might be forced north of Australia, through the Torres Strait that separates the continent from New Guinea.

By mid-August, as he neared his seventy-first birthday, Willis had given up trying to force the raft south and was now simply trying to keep from being smashed ashore in the New Hebrides islands. Small, steep-sided islands rose up out of the sea to challenge Willis' course. Threading his way through the Banks islands of Merig and Mera Lava, Willis found himself less than a mile from Merig's coconut-laden shore. The fresh food taunted Willis to the point where he could bear it no longer. He maneuvered his ten-foot Samoan outrigger into the water, jumped in, and paddled hurriedly toward the seas breaking on the rocks of Merig. *Age Unlimited* hovered unanchored and unmanned, but largely motionless on a calm sea.

It was a daring gamble by a seventy-one-year-old man, especially when he could find no safe route to shore. The heavily-laden coconut trees seemed to taunt him from shore. When a sudden squall made up, Willis turned to see his raft adrift. When he turned his canoe to give chase, the outrigger snapped its fittings, leaving him wallowing. It took an hour of excruciatingly delicate paddling before he brought himself alongside the raft. His flying expedition for fresh supplies a failure, Willis could take a small measure of comfort in his continued existence in the face of such a risk.

In early August, Willis charted a new course west, through the Coral Sea and towards the Great Barrier Reef of Australia. On the night of August 8th, while attempting to pull his jib over to the port side, a line gave 'way and Willis was knocked back against an iron boom. For the rest of the night, Willis could not move, and feared that he had cracked a vertebrae and might be paralyzed.

All the next day Willis lay crumpled on deck, as the raft drifted north on its own, towards the Solomon islands. With his legs useless, he had to drag himself along deck to the cabin to open a can of beans to feed himself. Four days later, with bruised and swollen black flesh hanging from his back, he managed to get to his feet and feel some strength return to his legs. Somehow, once again, he had survived an injury that would have swept most off the decks and into the sea. When a steamer hailed the raft on Willis' seventy-first birthday, August 19th, Willis merely told the captain to report his position to his wife, if you please.

By the turn of the month, Willis was approaching the Great Barrier Reef. He made the raft ready, placing lines where he could reach them in an emergency, fashioning a new anchor to replace the one washed overboard while he lay paralyzed, and cautiously searching through his binoculars for signs of the danger ahead. When the reef appeared, Willis steered the raft toward the widest opening he could find. As the sounds of the Pacific rolling into the continent crashed around him, *Age Unlimited* skimmed through a narrow channel in the reef. With the centerboards raised to miss protruding coral heads, the raft scudded to a standstill atop the reef as the tide washed out and left Willis high and dry. He lowered himself down from the raft and, with a small bucket in hand, went searching for edible invertebrates in the tide pools.

A returning tide lifted the raft free, and Willis was able to sail a short distance before being stuck fast once again. Another tide, and the raft washed free and Willis made haste to raise sail and head off for the northwest, only to be blown back onto the reef again. When he finally got the raft free, Willis headed directly west, finally clearing a path between Brook and Hitchinbrook islands off the coast of Queensland in Northeast Australia. Willis was coming ashore through waters first explored by James Cook two hundred years earlier. At eleven o'clock in the morning, September 9th, 1964, *Age Unlimited* approached a long low sandy beach. Willis tossed out his makeshift anchor, collected his passport and clearance papers from Samoa, and jumped into the waist-high surf. After a short search he saw a human being he could hail. "I'm Willis from New York (ibid, 244)," was all he said. After 11,000 miles and 204 days across the whole breadth of the Pacific, the journey from Peru was over.

Like many great voyages, the end was anticlimactic. Several shipping lines collaborated to bring the *Age Unlimited* to the U.S., where Willis donated it to the Mariners' Museum in Newport News, Virginia. Like his earlier raft, this one too would be forgotten, and within ten years the museum had discarded it as a rusting, unmanageable curiosity.

Chapter 5

Heyerdahl Takes on the Atlantic Ocean

Ra
Ra II

"One clings to [reeds], figuratively, when all else has failed."

—Thor Heyerdahl
The Ra Expeditions (1971)

Ra (1969)

Ever since his archaeological research on Easter Island in 1956-57, Thor Heyerdahl had been confronted by a dilemma. Norwegian archaeologist Arne Skjølsvold, one of the five archaeologists on the Easter Island dig, had excavated a unique moai statue, which was known locally as No. 263. It was special because a strange ship had been carved into its chest. And even though Skjølsvold (pers. comm, March 14, 2000) has come to believe that the carving probably represents a European sailing vessel—one cut into the stone after the statue was erected—one can also interpret it as a prehistoric reed vessel. So the distinct possibility existed that the *Kon-Tiki* experiment might have been conducted with the wrong type of prehistoric raft.

The carving appears to be propelled like a three-masted, square-rigged European ship, yet the hull has the distinctive upswept bow and stern of a Lake Titicaca reed ship. There is also a strange line trailing from the bow of the vessel, which some believe might be a turtle caught by one of the many sailors apparently standing on deck, and which others interpret as a kind of stone anchor.

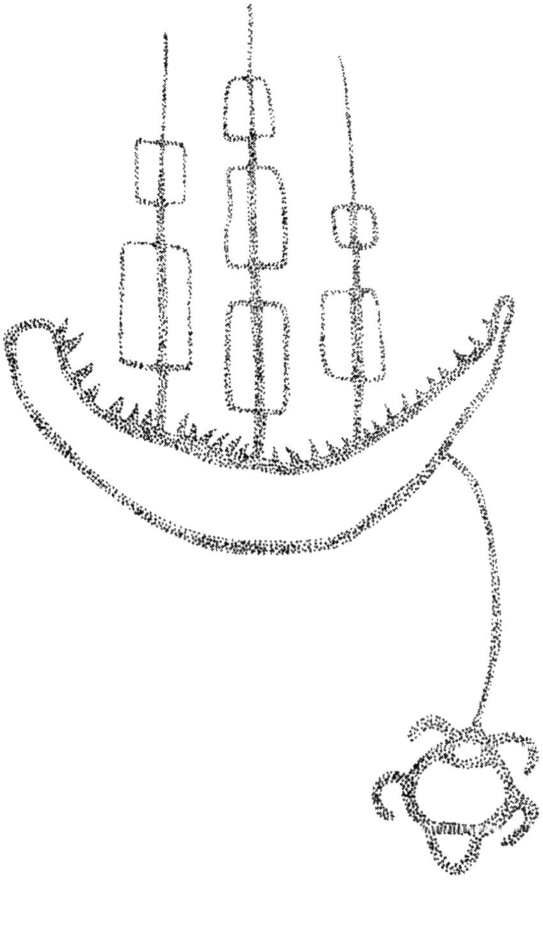

Fig. 11. The ship carved into moai no. 263 on Easter Island (after Heyerdahl 1961)

Beyond the carving on the chest of No. 263, Heyerdahl's expedition also uncovered locally produced volcanic stone carvings that unmistakably represent reed boats. Taken together, Heyerdahl saw this as evidence that reed ships had arrived on Easter Island at some point in prehistory, perhaps bringing the stone cutters who produced the giant moai or the finely hewn stone platforms on which they rested. If one accepted that a boat made essentially of grass could traverse large sectors of the Pacific, could one cross other oceans as well?

Heyerdahl began to research boats constructed of reeds all over the world, traveling from museum to marsh to study watercraft made from totora, papyrus, and other natural fibre materials. When he saw boats and barges apparently constructed of papyrus depicted in the burial chambers of Egypt, his experimental instincts took over. If a reed boat had carried the legendary *Kon-Tiki* across the Pacific 1,500 years ago, could a vessel built of papyrus have carried a crew of Egyptians across the Atlantic 5,000 years ago? It was not a question Thor Heyerdahl could formulate without attempting to provide an answer.

Papyrus was scarce in Egypt, and Heyerdahl had his own doubts that it possessed the same watertight capacity of totora reed. Papyrus experts brought in to advise the Egyptian government on Heyerdahl's request to use a plot of land near the Great Pyramids of Giza for the construction of a papyrus boat were uniformly hostile. None thought papyrus could remain afloat for much more than two weeks, and that only in still, fresh water. No one thought such a vessel could survive three thousand miles of a rough transatlantic crossing. When one of the experts announced that laboratory tests had shown that pieces of papyrus sank after a few days, Heyerdahl thought to himself that this was like throwing a piece of iron into the sea and deciding that any ship built of iron would automatically sink as well.

When the Egyptian authorities finally gave the go-ahead for a papyrus ship to be built at Giza, Heyerdahl was suddenly confronted with the enormity of the experiment he had set for himself. The complexities of the project and their financial burdens were almost overwhelming. Heyerdahl selected Buduma tribesmen from Chad, who still lived on floating islands of reed, to build his transatlantic ship. With a sketch by the Swedish scholar Björn Landström of what an ancient Egyptian papyrus vessel should look like, Heyerdahl then required 300,000 papyrus stems to be harvested from Lake Tana in Ethiopia, then transported to Egypt across the Red Sea and through a war raging around the Suez Canal. He had to assemble a crew and build a vessel no one had built in 5,000 years, then transport both vessel and crew to Morocco and be ready to sail in six months. As Heyerdahl himself wrote:

> ...the reeds must be properly sun-dried or they would rot in their bundles... In fact, the inflammable reed must be landed in Suez and transported along blocked roads to rejoin the Nile near Cairo. Before the papyrus reached the pyramids, a camp with all conveniences including a cook and provisions must be ready in the desert for the necessary guards and labor force... Sails and rigging, ancient Egyptian steering mechanism, wickerwork cabin, specially made earthenware storage jars and ship's food prepared as in ancient times... (Heyerdahl 1971, 105).

As construction of the replica progressed, Heyerdahl divided his time between the work site at Giza and trips to the tombs to study frescoes showing representations of ancient Egyptian watercraft. A peculiar cable running from the stern of a reed ship to its afterdeck caused particular consternation, as no one could provide a satisfactory explanation for its presence. Did it simply hold the shape of the inwardly curved reed stern, or did it have some more significant maritime function. Heyerdahl's boat-builders from Chad eventually did away with the cable once they had curved the papyrus stern into shape, dismissing it as unnecessary.

Fig. 12. Sketch by Swedish maritime history expert Björn Landström, upon which Thor Heyerdahl based the construction of the reed boat *Ra* (after Heyerdahl 1971)

In the meantime, Heyerdahl canvassed the globe in search of a multi-national crew that would show the world that men from different nations could in fact work together toward a common goal, despite a constant stream of news reports to the contrary. In the end, Heyerdahl put together a crew consisting of himself and one of his Chadian boat builders, an Egyptian diver, an Italian mountaineer, a Mexican anthropologist, a Russian doctor, and an American navigator.

In New York, Heyerdahl met up with that young navigator, named Norman Baker, whom he had met in Tahiti back in 1957. As Baker remembered the meeting, the world-famous explorer was proposing that he, Baker, navigate a boat made out of paper across several thousand miles of ocean. Baker had no problems with the navigation. As he saw it, any novice should be able to drift away from Africa and fetch up somewhere in America. If not, he should turn in his certifications. But to navigate a boat made out of paper? That was something else again.

> "When Thor came to New York, a mutual acquaintance asked if I wanted to meet him, and of course I had already met him in Tahiti. But obviously I wanted to meet Thor Heyerdahl again, so we met and had dinner. It was then that he began to talk about crossing the Atlantic on a reed boat. Oh my gosh. I thought he had finally gone off the deep end. Now, I could well understand crossing an ocean on a boat made of logs the size of telephone poles. But going across an ocean on a boat made of reeds? Grass? Straw? I said 'C'mon, Thor.'" (Norman Baker, personal interview, 8 July 2000)

Yet the dinner between Baker and Heyerdahl lasted four hours. By the end of it, Baker was convinced by Heyerdahl's arguments that ancient mariners had in fact crossed prehistoric waterways on boats made essentially from grass. This was no casual experiment. Heyerdahl had invested years of global research in studying the construction and use of modern reed boats, as well as the records and paintings that pointed to their use in prehistory. Heyerdahl presented Baker with sketches of one hundred and thirty-five archaeological similarities between ancient Mediterranean civilizations and prehistoric American civilizations. "Personally," Heyerdahl confided to Baker, "I am quite sure contact was made. But I am not believed; in fact, I am ridiculed even more than before *Kon-Tiki* because, after all, I am proposing that the mechanism of contact was a grass boat." (ibid)

Heyerdahl proposed a test of the reed boat, not to demonstrate that contact had in fact occurred, but to show that a valid mechanism for such contact existed. Such experimental niceties, however, went unnoticed by most commentators. When he arrived in Egypt as a member of Heyerdahl's crew, Baker overheard Heyerdahl stressing to a reporter that he was not trying to show that Africans had settled the New World, only to assess the maritime technology available to sailors 5,000 years ago. No sooner had Heyerdahl departed than the reporter rushed to a phone, called his bureau, and shouted "Thor Heyerdahl is trying to prove that America was populated by the Egyptians!" (Baker 1997)

Despite the obstacles thrown in his path by logistics and critics alike, Heyerdahl had his raft ready to be delivered to the African coast on April 28, 1969, the twenty-second anniversary of the start of the *Kon-Tiki* expedition. He named her *Ra* after the Egyptian sun god. A month later, the papyrus boat sat absorbing seawater in the Moroccan port of Safi. For eight days, as the raft underwent last-minute rigging and provisioning, it continually lost precious hours of its brief projected life span. Finally, on May 25, Heyerdahl ordered the paper boat towed to sea.

Just as on the *Kon-Tiki*, the crew marveled at how waves washed through rather than swamped the hull. Norman Baker found himself incapacitated by influenza and a temperature of 102°. As the raft fought to make its way offshore on the very first day at sea, Heyerdahl was suddenly called aft to see a disaster in the making. Not one but both of the steering oars had snapped completely away. The crew was despondent, feeling that the experiment was over before it had even begun.

Heyerdahl studied the raft closely. Without the two oars, the raft suddenly turned on its own and headed out to sea. Heyerdahl was concerned but not despondent. For a man who had spent his life trying to demonstrate that ocean currents were conveyors of culture instead of impassable barriers to human migration, *Ra* had suddenly turned into an even more daring experiment. It was one of Heyerdahl's favorite scenarios: ancient mariners on a crippled vessel, its crew in a battle against wind and current, intent on carrying their culture to a foreign shore. *Ra* would now imitate the navigation of Incan rafts with their guara leeboards.

Hearing Heyerdahl yelling, Norman Baker crawled from the wickerwork cabin to learn the good news. He was stunned to learn that the vessel of which he was navigator was now adrift, without rudders, and almost completely unmanageable. The raft broached and waves slammed its sides. At a stroke, the crew had not exactly been transformed from sailors into supercargo, but Heyerdahl's Atlantic laboratory had subjected his raft experiment to its first variable.

The next challenge for the raft was to pass the lowlands at Cape Juby on the African coast, and begin to arc to the west, across the Atlantic. To everyone's surprise, the raft began to list slowly toward the wind, not against it, as on a regular sailboat. Heyerdahl discovered that waves breaking on the windward side deposited water more heavily into the reeds on that side of the raft, while the lee side remained high and dry. Much shifting of cargo failed to alleviate the problem, which was yet another lesson relearned after five millennia in obscurity. On May 31st, *Ra* passed Cape Juby and put Africa astern. The reed boat had been afloat two weeks, without showing any signs of the imminent disintegration predicted for it. But the greatest challenge still lay ahead.

The crew improvised a new steering oar from a spare mast, being somewhat abashed that they had almost been reduced to passengers. They might have wished later that they had been. Norman Baker brought along a small box of books and never opened a single one. The crew found itself working day and night, seven days a week, for two straight months.

The raft moved westwards at approximately sixty miles per day, half again as fast as Tim Severin's bamboo raft *Hsu Fu* would snake across the North Pacific nearly a quarter century later. But where Severin's wash-through bamboo raft subjected its crew to constant drenching, Heyerdahl's paper boat, at least for the time being, kept his men high and dry.

> It was unspeakably good to crawl to rest inside a warm sleeping bag. You woke up with such an infernally good appetite. You felt an extraordinary physical well-being. Small pleasures grew big; big problems felt small. The Stone Age life was certainly not to be despised. There was no reason to believe that people who lived before us, using their bodies strenuously, merely endured hardship and never received their share of life's joys (ibid, 195).

The performance of the special rigging, copied so carefully from Egyptian tomb paintings, now seemed in practice to both Heyerdahl and Baker to have been designed to handle ocean swells and waves. It seemed clear enough that the papyrus boats of antiquity had done more than float calmly on the Nile. However, the absence of the strange cable linking the upturned stern with the aft deck—the cable that the boat-builders from Chad had told Heyerdahl was unnecessary—now came back to haunt the raft. Halfway through June, the stern of the raft was awash, even as the bow continued to sail on as dry as the day the reed boat was launched.

By the first days of July, the *Ra* had sailed over 2,000 miles, and less than 1,500 remained between the raft and the Caribbean island of Barbados. Heyerdahl was learning that, unlike the pre-Incan log raft *Kon-Tiki*, this Egyptian reed ship required true sailing ability. Anybody could hang onto a raft. The stern continued to drag further into the sea, the starboard side continued to fill with water. The oars had been rebuilt only to snap again.

On Ra, Heyerdahl felt as if he was driving a car without a license. It was the fate of anyone who tried to recreate prehistoric technology. A simple and many would claim unbridgeable flaw was always present in the experiment. The replica might be correct; the ocean route might be the right one. But no scientist could replicate the prehistoric mind, nor venture with certainty into a prehistoric worldview. As the Caribbean loomed just beyond the horizon, the *Ra* slipped inexorably into the sea.

On July 18th, 1969, with sharks and Portuguese Man-o-war encircling the reed boat, Heyerdahl ordered the crew to abandon ship. The apparent failure of this greatest of primitive reed ship experiments was wholly overshadowed in the days that followed by the first landing of humans on the surface of the moon. The juxtaposition of the sinking trajectory of the ancient raft with the successful soft landing of the ultramodern lunar module could not have been more striking. Between the failure of *Ra* and the triumph of Apollo 11, it seemed as if modern humans had put the ancient world behind them once and for all. All modern humans, that is, except Thor Heyerdahl.

Ra II (1970)

In the spring of 1970, under a thick blanket of secrecy, Thor Heyerdahl was back in Morocco, building another ship of papyrus. This time, Heyerdahl had gone back to his roots, bringing Aymara Indians from Lake Titicaca to Africa to oversee the construction of perhaps the most beautiful recreation of a prehistoric vessel ever undertaken. With the many lessons learned from what was now called *Ra I*, along with another year of research amongst traditional reed boat builders, Heyerdahl left nothing to chance this second time around. Even so, he was plagued by worries that a second failure would prove disastrous, and be seen as "nothing but a risky repetition" (Heyerdahl 1971, 286).

After setting sail for Morocco on May 17, 1970, the *Ra II* proved less stable that her predecessor, but more seaworthy. The new reed ship was thirty-nine feet long—almost twenty feet shorter than *Ra I*—sixteen feet wide amidships, and six feet deep. In fact, the raft proved seaworthy enough to transport a multinational crew of eight completely across the Atlantic. But the voyage nearly ended less than a month after it started.

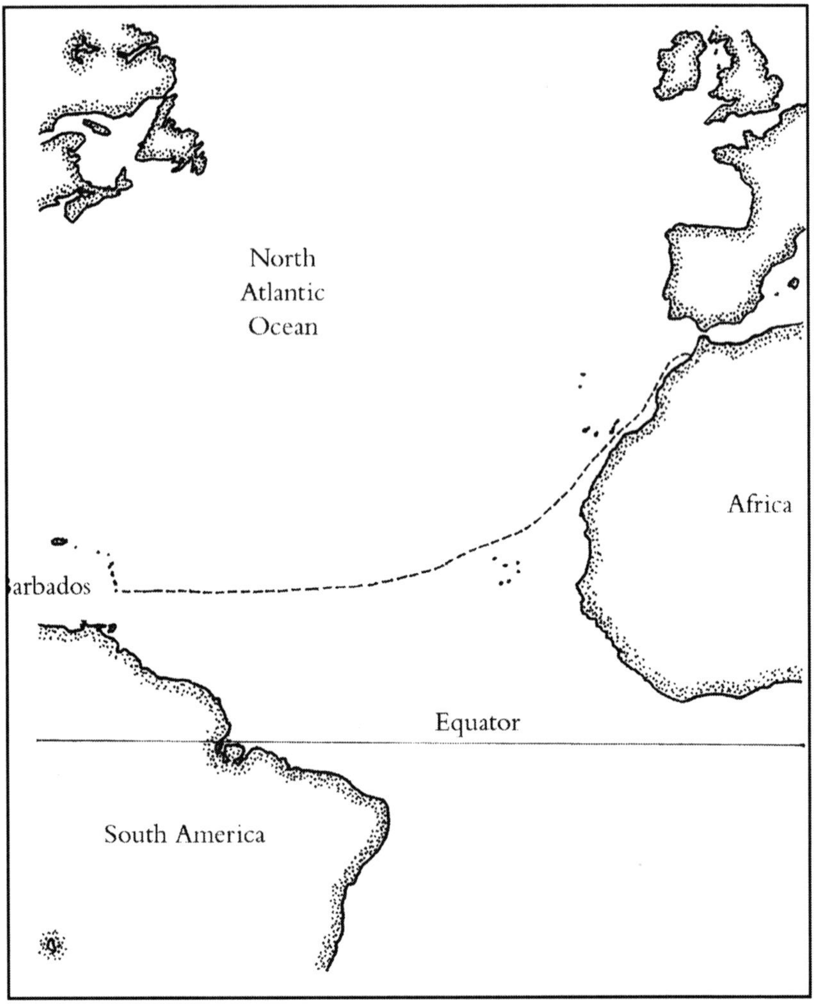

Fig. 13. Route of *Ra* and *Ra II* across the Atlantic (after Johansen 1999)

Norman Baker had returned as well, as both navigator and second-in-command. As Baker set a course away from Africa as fast as possible, none save Heyerdahl were very optimistic about their chances for success. Leaving port, the reed deck of *Ra II* was only three feet above the surface of the sea. Within two weeks, the raft had sunk two whole feet. After three, the decks were awash. Seeing this, Baker thought they had little choice but to run for the Cape Verde Islands, 1,200 miles from Safi in Morocco. There the expedition could slink into obscurity, rather than cause a scene by calling for an inevitable rescue in mid-Atlantic.

Desperate, the crew threw over the sides everything they could, even to the point of potentially sacrificing themselves by throwing their papyrus life raft overboard. As the Cape Verde Islands drew closer, so did the crew's last apparent hope of abandoning ship before it sank. They had all decided that they would sail into port and go home. But as Baker took the raft's position throughout that day, Heyerdahl kept a steady course westward. In the morning, the Cape Verdes were a few points off the port bow; by noon, they were directly abeam, about eight miles away; by early evening, the islands were on the port quarter, about sixteen miles away. The raft had passed the point of no return.

No one said anything. The raft had less than a foot of freeboard remaining. There was no way to turn the reed boat around and return the Cape Verdes. The crew knew that they were now committed to 2,000 miles of open ocean. To solidify their dismay, the crew decided to conduct a secret ballot on the only question that mattered: would *Ra II* make it across the Atlantic Ocean? As Baker recalled, seven of the crew voted "no"; only one responded "yes." "No one ever asked who the cockeyed optimist was," Baker remembered, although in the end it was the lone dissenter who proved to be correct. "Though I've never asked, in my heart I know who it was—our Captain, Thor Heyerdahl." (Baker 1997)

Still, as the raft somehow passed the halfway point on its voyage, the crew underwent a remarkable transformation. The depression of being on a sinking raft at the beginning of the expedition suddenly became an equally dangerous manic optimism that success was now a foregone conclusion. A few days later, an enormous wave smashed into the *Ra II*, and the steering oar, as big as a telephone pole, snapped. Once again, a reed boat was adrift in mid-Atlantic. Worse, the materials put on board to cope with such an emergency had been pitched overboard in the previous attempt to lighten the raft.

Later that night, after an exhausting day of trying to repair the steering oar, the crew had slipped back into their earlier depression. Baker asked about that night's dinner. Santiago Genovés offered crackers and cheese. Baker said "No. We're going to have a hot meal to keep up our strength." Baker realized that their morale was slipping. If their strength waned as well, they were as good as dead. Again the Mexican balked, too tired to cook. "We're having a hot meal tonight if I have to cook," exploded Baker, "and I'm not a cook." It was the only time on either voyage that Baker had vociferously countermanded anyone's opinion. It was then that the six-foot five inch Egyptian, Georges Sourial, gently pushed Baker aside and said, "Norman, you don't even know how to cook. Everybody aft! I will cook" (Norman Baker, personal interview, July 8, 2000). With hot food in their bellies, morale improved, and the raft drifted westwards, inexorably closing the gap with the Western Hemisphere.

The ancient vessel drifted through a sea of modern refuse. Lumps of oils, large and small, dogged their path. The sea was so polluted that it appeared as if it might dissolve the raft itself. Baker oversaw an experiment within the experiment to tilt the double mast forward to gain additional drive from the sail and improve the steering.

On July 12, 1970, the crew sighted land. They had sailed 3,270 miles from Africa. As the reeds began to lose their buoyancy once and for all, *Ra II* sailed hard into the harbor at Bridgetown, Barbados. Heyerdahl had staked everything on a second reed boat, and it had delivered him to the New World. Soon after, *Ra II* was returned to the Old World, and took up a permanent place in a great hall in Oslo, directly adjacent to that occupied by *Kon-Tiki*.

In the years following the *Ra* expeditions, Thor Heyerdahl continued to study the reed boat construction methods of traditional peoples. His focus shifted from Egypt to Mesopotamia, the cradle of civilization. In the British Museum, he saw a wall relief removed from the famous archaeological site and ancient city of Nineveh, across the Tigris River from present-day Mosul in Iraq. The relief apparently depicted a sea battle between two reed ships thousands of years ago. Yet modern scholarship barely mentioned reed ships in ancient Mesopotamia. Heyerdahl noticed a jar in the museum as well, one that seemed to indicate that ancient reed boats were fitted with keels or centerboards. If true, it was an apparent indication that this vital technological artifact was not exclusive to Asia and Ecuador as scholars had long presumed.

Heyerdahl recognized at once that if Sumerian or Assyrian cultures in fact used reed ships for naval battles, and for more than river travel, they must have known how to sail them, not simply drift on board them. If Heyerdahl wanted to retrace an ancient sea route down the Tigris-Euphrates Valley, then beyond the Persian Gulf and into the Indian Ocean, he would not have ocean currents at his disposal. For the first time in his career as history's foremost archaeological experimenter, he would have to reinvent a way of sailing that had not been used in thousands of year.

Chapter 6

Heyerdahl's Last Great Experiment

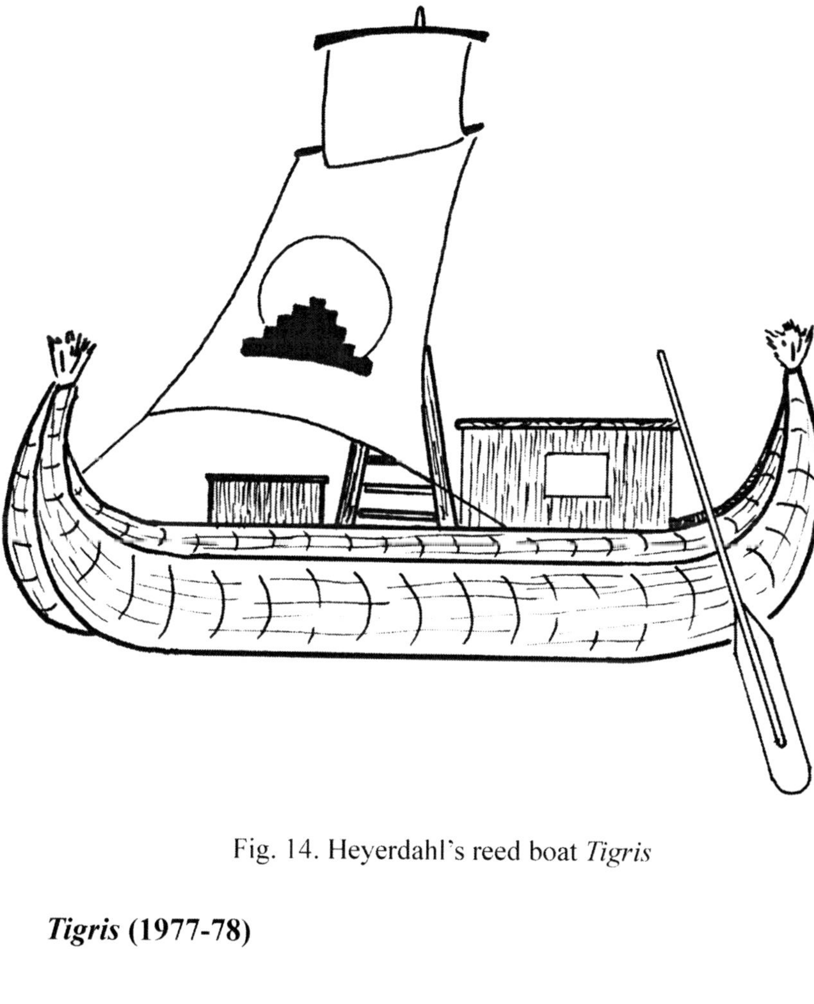

Fig. 14. Heyerdahl's reed boat *Tigris*

***Tigris* (1977-78)**

In Iraq itself, Heyerdahl gathered evidence for the ancient use of maritime vessels made from a tall freshwater reed called berdi. There was more information, as well, on coatings the ancients may have used to protect these reeds against water absorption. Asphalt in some kind of mixture with pitch and oil was mentioned. Much more importantly, Heyerdahl learned from the marsh Arabs of Iraq a vital piece of data on the performance of berdi that would influence the entire outcome of his planned experiment. Berdi, they said, must be cut in August, and only in August, or it absorbs water quickly and sinks. The berdi cut in August was dried for two or three weeks and then used for the reed houses in which the Arabs dwelled. Estimations of the buoyancy of properly harvested berdi ranged to upwards of a year. This was a new, seasonal aspect of reed boat construction that no one had considered before.

As with *Ra II*, Heyerdahl brought Aymara Indians from Lake Titicaca to construct his new reed ship in Iraq in September 1977. When Norman Baker arrived to join the expedition, he was impressed with model tests of the raft. Unlike the previous reed boat expeditions, the *Tigris* would be subjected to severe tests of both river and coastwise navigation. But the crew was supremely confident as the construction progressed. As Norman Baker recalled, by the time of the *Tigris* expedition, Heyerdahl had unlocked the secret to reed boat construction that he had been searching for for fifteen years.

> "These boats could be made indefinitely buoyant if the reeds were harvested at the right time, in the month of August, when the sap was in the entire stalk—a stalk which grows more than fifteen feet high. It's only when the blossom blooms at the top of the reed, that it is filled with sap. After that bloom is over, the sap retreats back down the reed stalk to the roots. We had harvested these reeds with their capillary tubes completely empty, in December, on *Ra I* and *Ra II*. The reeds for *Tigris* were not only harvested at the right time of year, the boat itself was fitted with a centerboard, as Thor had seen on the

jar in the British Museum." (Norman Baker, personal interview, July 8, 2000)

These two improvements led to the most seaworthy reed ship ever constructed in the modern age. When completed, this new reed ship, christened *Tigris*, was sixty-feet long, nearly twice the length of *Ra II*. To handle the bigger craft there were eleven crewmembers instead of the previous eight. On November 11th, the *Tigris* was ready to slide into the waters of the Shatt-al-Arab waterway, the gateway to the Persian Gulf. But when the massive reed ship was maneuvered toward the river, the bow tilted into the water as the stern stuck in the mud. It was a near-disaster, especially since Heyerdahl's Aymaras had already returned to South American, and if the ship were badly damaged there were no expert reed boat builders on hand to repair the vessel. A passing Russian truck later supplied the necessary shove to get the *Tigris* afloat, and the next day snorkelers reported to Heyerdahl that no serious damage had been done.

The next challenge was to get the cumbersome craft downriver to the sea. But the river currents waffled the ship from shore to shore, as a chase boat scurried along behind. The massive steering oars dug into the river bottom, threatening to snap. Then the river smoothed out, and the crew began to gain in their ability to control the vessel's movements. They spent their first night on board more confident in both the *Tigris* and themselves.

During the days that followed, the *Tigris* moved slowly down the heavily polluted Shatt-al-Arab waterway, with cakes of white chemical broth seeping along the edges of the berdi reeds. From the air, it looked like the reed ship was sailing through the ice floes of the Arctic Ocean. Extracting the reed ship from the Gulf turned into an even greater challenge. Contrary winds and currents, looming oil platforms and supertankers, and a lack of knowledge of how to turn the great ship into the wind, forced *Tigris* into dangerous shallows, from where Heyerdahl was forced to pay ransom for a tow by a pirate dhow. Later, a Russian ship towed

Tigris toward safer waters near Bahrain, but not before ripping large chunks of the berdi reeds from the bow of the ship.

Landing at Bahrain, Heyerdahl searched for the legendary site of Dilmun. It was this legendary 'place in the east' from which Sumerian mariners claimed their origins, and to which they apparently returned to gather raw materials from local mines. Heyerdahl and Norman Baker studied dock areas where ancient mariners loaded blocks of stone onto shallow draft vessels more than 4,000 years ago, and speculated how this could have been accomplished with a reed boat like the *Tigris*.

Once safely outside the Straits of Hormuz, Baker and the crew began to gain control of *Tigris*. The reed ship now responded well to the tiller and, following the curve of the Arabian peninsula, the whole of the Indian Ocean lay before the crew. First, *Tigris* called at Oman, where a Mesopotamian ziggurat-style pyramid—the first located outside the Tigris-Euphrates valley itself—had just been discovered. Here, Heyerdahl once again searched for the fabled site of Dilmun.

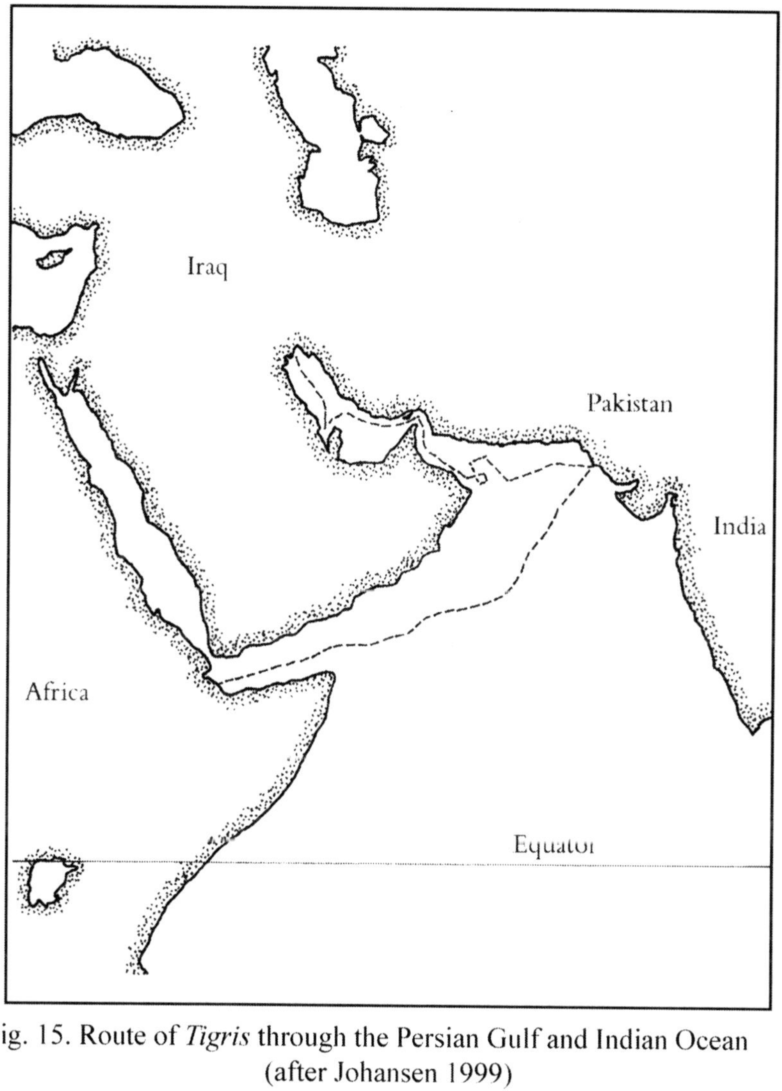

Fig. 15. Route of *Tigris* through the Persian Gulf and Indian Ocean (after Johansen 1999)

As 1977 turned to 1978, Heyerdahl set course for Pakistan, and the ancient culture center of the Indus valley. On January 26th, 1978, the crew of *Tigris* picked up the island of Astola off Pakistan. Heyerdahl had demonstrated that a primitive reed boat could link Mesopotamia with the Asian sub-continent civilization of the Indus Valley. If the experiment had ended here it would have been considered an enormous success, but Heyerdahl wanted to go even farther. As he wrote: "We were learning from people with centuries of experience, and were at any rate doing far better than during the first fumbling experiment with Ra" (Heyerdahl 1980, 261).

With the reeds floating higher after three months than *Ra II* did after three weeks, the crew agreed to cross the Indian Ocean once again, and try to reach the edge of the Egyptian realm near present-day Somalia. Heyerdahl thought they might even sail down the coast of Africa and cross the Atlantic to the Western Hemisphere, and demonstrate the global reach of the reed ship. With finances dwindling, the crew instead set sail for Somalia, in an attempt to link the three great culture areas of the Old World.

Two months later, on March 28th, the *Tigris* raised the African coast. But there was nowhere to land safely. All of the small nations around Djibouti seemed to be at war with each other. Baker and several of the crew were in favor of sailing *Tigris* up into the Red Sea, but neither nation on its shores responded to Heyerdahl's request. Military operations filled the waters and the skies. It was easy to conclude that humanity had made little progress since the last reed boat arrived from Mesopotamia 4,000 years ago. Frustrated, Heyerdahl called the crew together, where they drafted an appeal to the United Nations, protesting "against the inhuman elements in the world of 1978 to which we have come back as we reach land from the open sea" (ibid, 336). The crew decided to burn *Tigris* rather than leave it to rot.

Miserably, they watched their primitive home of five months go up in flames. Yet Heyerdahl and the three of his *Tigris* crew who had also been with him on *Ra I* and *Ra II* had little reason for remorse. They had now sailed more than 10,000 miles on board reed boats, a total that put together would amount to a drift halfway around the world. These three ships—two of papyrus and one of berdi—had been constructed from materials which, only a decade before, every expert had felt certain could not last more than two weeks on the open ocean. Like a dramatic maritime funeral of a Viking chief, the burning of *Tigris* was a fitting conclusion to Thor Heyerdahl's career as the greatest transoceanic raft expedition leader in history.

CHAPTER 7

CLOSING THE ARGUMENT: CHILE TO EASTER ISLAND

> Above and below, the forests on his side of the river came down to the water in a serried multitude of tall, immense trees towering in a great spread of twisted boughs above the thick undergrowth; great, solid trees, looking sombre, severe, and malevolently stolid, like a giant crowd of pitiless enemies pressing round silently to witness his slow agony. He was alone, small, crushed. He thought of escape—of something to be done. What? A raft! He imagined himself working at it, feverishly, desperately; cutting down trees, fastening the logs together and then drifting down with the current, down to the sea into the straits.
>
> —Joseph Conrad
> *The Outcast of the Islands*

Viracocha (2000)

By the time of the 50th anniversary of the *Kon-Tiki* expedition—the spring of 1997—no primitive raft had made the connection between South America and that continent's closest Polynesian island, although Eric de Bisschop had come close in *Tahiti-Nui I* in 1957, and Kitin Muñoz had drifted on *Uru* from Peru to the Marquesas in 1988. Thor Heyerdahl's massive American Indians in the Pacific had compiled a much greater case for prehistoric reed boat—as opposed to balsa raft—voyages to Easter Island, yet no one could say from actual experience whether such an expedition was possible.

It was by far the most interesting of all possible potential replica raft expeditions. It would connect the reed boat builders of the pre-Incan highlands of Peru with the Pacific ports that could have made use of large versions of lake reed boats. It would connect a potential prehistoric South American port with the most likely Polynesian island to have received cultural impulses from pre-Incan Peru. It would offer a plausible escape route for the legendary *Viracocha*, as he fled from the shores of the great highland lake to his exile somewhere toward the setting sun. And it would offer some comparative strength to Heyerdahl's hypothesis that Moche mariners forced to sea during cultural catastrophes triggered by El Niño events would be pushed southwest toward Easter Island. Little did one realize at the time that an American explorer was laying the groundwork for just such an expedition.

Just two years later, an explorer from Massachusetts proposed the most ambitious reed boat expedition ever attempted. Phil Buck, a highly-accomplished mountaineer who at the age of thirty-six had already amassed more exploration firsts than most adventurers gather in a lifetime, announced that he would circumnavigate the globe in a series of five reed boats. The reed ships would all be named for the bearded sun-god Viracocha. The first leg of this enormous undertaking would take the *Viracocha I* from Arica, a Chilean port near the border with Peru, to Easter Island.

"I had dreamed of a voyage by reed boat after reading Thor Heyerdahl's fantastic books at the age of eleven, twenty-five years ago. I started seriously thinking about a trip from Arica to Easter Island in 1995 because I thought that connecting the Andes with Easter Island was the most important voyage that could be done at the time. And I understand why it had not been attempted, for most thought it to be impossible. If we had reached any other island besides Easter Island, I myself would have considered our voyage unsuccessful" (Phil Buck, personal communication, August 14, 2000).

In the late 1998, Buck journeyed to Huatajata, Bolivia, on the shores of Lake Titicaca. There, both the Catari and Limarchi families—many of whom had helped Heyerdahl construct both *Ra II* and *Tigris*—collaborated with Buck to build the hull of *Viracocha I*. The hull was completed in March of 1999, and then stored under a tin roof until it was trucked to the coast in December. During the storage period Buck had the boat builders haul on the hull's ropes each month. Eventually, he considered that this simple step might have been critical to the success of his voyage. Reeds shrink as they dry, and having a solid reed boat is imperative to lower water absorption and lessen the overall flexing of the boat at sea.

The hull was trucked to Arica, Chile, in December 1999, where it was fitted out with masts and sails, rudder oars, and a bamboo cabin. Like Heyerdahl, Buck selected a multi-national crew, in this case three Chileans, a Bolivian, a Britisher, a Frenchman, and one other American.

Buck envisioned a six-week expedition to Easter Island, one that would start around January 15 in the new millennium, and reach Easter Island in early March. There the explorers would tramp the island, look in on the totora reeds of the island's crater lakes, and seek permissions for the second leg in the global circumnavigation, from Easter Island to Australia. In the end, Buck's timetable evolved with almost clock-like precision. *Viracocha I* left Arica on February 25, 2000, after having been placed in the water fifteen days earlier. Norman Baker had suggested to Buck that the raft should be put into the ocean three days prior to departure, to allow the reeds to absorb seawater as ballast, but problems with inspections and computer software delayed the departure.

Three days after leaving the Chilean coast, the reed boat was 150 miles to sea, sailing 150 degrees off the wind and about two knots an hour, on a southwest course toward Easter Island. Over the course of the next forty-four days, the reed boat continued to average about two knots an hour, or about fifty miles a day. On only two or three days did the boat's performance increase or decrease dramatically. On March 9, contrary winds stalled the boat's progress to nothing, while on March 23rd and 26th, the boat skimmed along with the wind at speeds of five and six knots.

By April 1st, 2000, the *Viracocha* I had sailed more than 2,000 miles from the coast of Chile, and was only 300 miles from Easter Island. As he neared the island of Sala Y Gomez, an uninhabited island 240 miles from Easter Island, the raft began to experience major wind shifts and velocity drops. The winds began to move in a counter-clockwise direction and at time would completely drop off. It was in stark contract to what Buck had experienced during the first month at sea, when he had steady trade winds from the South and South-East. The raft drifted on calm seas six miles from Sala y Gomez. Buck was now within a long stone's throw of an island Heyerdahl landed on in 1956. He was well aware of the significance of his achievement:

"I thought that passing near Sala Y Gomez was important to Thor Heyerdahl's theories because passing an uninhabited, rocky bird island was one of things that the Spanish Chronicler Sarmiento had heard that Inca and Pre-Incan voyages would pass one week before reaching the island Thor Heyerdahl believes to be Easter Island" (ibid.).

Buck hoped to zero in on the harbor at Anakena, the port on the northeast corner of Easter Island where legends place the arrival of Hotu Matua, the prehistoric maritime explorer and founder of Easter Island culture. As the island emerged out of the Pacific a week later, Buck steered the reed boat around the southern side of the island, to a landing at Hango Piko.

For perhaps the first time ever—or possibly the first time in over a thousand years, depending on your view of the evidence—a reed boat sailed through the gap separating the ancient ceremonial center of Orongo, and the small offshore island of Moto Nui. It was across this gap that the famous annual birdman competition was held, where young men on small reed floats swam and climbed to reach the first sooty tern egg of each spring.

At 3:00 in the afternoon of Sunday, April 9th, 2000, *Viracocha I* anchored at Hanga Piko, where the arrival of a reed boat from Chile was greeted with decidedly mixed feelings. The memory of nineteenth century raids by Peruvian slavers is still warm on this Polynesian island, and any reminders of ancient ties with the mainland stir complex emotions. Buck himself wanted to preserve *Viracocha I,* which had survived the voyage in almost perfect condition. Local merchants likewise saw an interesting tourist attraction. But when others recalled the biting mosquitos they felt had arrived with the reeds of Kitin Muñoz' *Mata Rangi I*, Buck had no choice but to agree to burn the ship. Less than two weeks later, stripped of all useable equipment, *Viracocha I* went up in flames at Hanga Roa, an event that in itself may reflect the fate of earlier voyagers to this strangest of all the world's islands.

Chapter 8

The Way to Central America

Illa-Tiki
La Manteña
La Enduriencia

Illa-Tiki (1997)

Toward the end of the 1980s, a twenty-six-year-old Dallas, Texas businessman named John Haslett had earned enough money from speculating in currency futures that he suddenly 'retired.' A year later, with his money running out and with little desire to return to work, Haslett wandered into a used bookstore and walked out with a copy of *Kon-Tiki*. As a sailor and an avocational historian with an interest in the archaeology of Central and South America, Haslett became fascinated by the *Kon-Tiki* expedition. His father had always allowed his two sons the latitude to attempt risky adventures—for John it was canoeing a class five river at the age of eleven and his first parachute jump at age fourteen. Building a raft and sailing into prehistory suddenly seemed like an appropriate extension to his boyhood explorations. Haslett became determined to follow Thor Heyerdahl across the Pacific, by building a balsa raft and sailing it from Mexico to Hawai'i.

By the spring of 1994, Haslett had formulated his plans to such an extent that he felt he needed to seek the advice of Thor Heyerdahl himself. The Norwegian responded by stating his belief that the Hawai'ian islands were well within the range of a balsa raft. But the success of such a voyage depended heavily on the effective use of the guara, which Heyerdahl himself had demonstrated in experiments conducted five years after *Kon-Tiki.* To solve the problem of crossing the doldrums, Heyerdahl suggested that the raft could be paddled, depending of course on the size of the raft and the number of the crew. Heyerdahl reminded Haslett that for two hundred years Spanish caravels crossed the Pacific from Mexico directly to the Philippines, but always had to return on the Japan Current that carried them north of the Hawai'ian islands.

In addition to his contact with Heyerdahl, Haslett had been conducting research on a group of cultures occupying pre-Columbian Ecuador and known collectively as the Manteños. It was in this Manteño area of coastal Ecuador where—along with the maritime culture areas of Taiwan—archaeologists had placed the development of the guara centerboard. But it was the adventure of drifting across an ocean that drew Haslett to South America more than any desire to test an archaeological hypothesis.

Like Gene Savoy twenty-five years before, Haslett was intrigued by archaeological evidence linking the Manteño areas with those of the Aztecs of Mexico. Similarities between culture areas of Western Mexico and Ecuador included ceramics, clothing and, especially, techniques of metal production. Savoy believed that the two areas had been joined in prehistory by totora reed boats, and sought to demonstrate such a link in the awkward voyage of the *Feathered Serpent* in 1969. Haslett considered it much more likely that the giant balsa raft freighter, with its massive cargo of trade goods, made a more suitable vehicle for such contact. To demonstrate that a balsa raft from Ecuador could reach Mexico, however, one could not simply drift with prevailing currents. The raft would have to be sailed like any sailboat.

As Savoy had discovered, the northwestern coast of South America is a difficult place for primitive craft. Winds and currents play havoc with coastwise navigation. The waters are filled with rocks and reefs; surface and shore with pirates and bandits. If the raft lost control near the shore, a wreck was certain. If one took a course too far offshore, the raft would get caught in the Humboldt Current and be carried 4,000 miles into the Central Pacific. Then came the real challenge: to cross the Equator and the Gulf of Panama without getting stuck in the endless circulations of the Equatorial Counter Current, the same ocean current in which Vital Alsar's balsa raft *La Pacifica* was stranded for over one hundred days in 1966-67.

If he reached the coast of Mexico, Haslett would complete a voyage that no other primitive replica raft had ever navigated. Even so, with the fiftieth anniversary of the *Kon-Tiki* expedition only a few years away, Haslett did not feel that this would be accomplishment enough. He felt that if he reached Mexico he could then launch a personal adventure: to reach Hawai'i by using the prevailing winds and currents that would then be at his back. If successful, the whole dual experiment would offer both an experimentally demonstrated mechanism for contact between Central and South America, and a speculative route to the Hawai'ian islands for prehistoric mariners from the Americas. While the latter voyage was considered largely outside the scope of prehistoric possibility, the former was of primary interest to scholars examining maritime interactions between traditional cultures of Central and South America.

Thus fortified with equal parts information and inspiration, Haslett set out for Ecuador in January of 1995. There he managed to harvest nine huge balsa logs, and had them shipped by flat bed truck to the coastal village of Salango. The twenty-ton raft, an almost precise copy of Heyerdahl's *Kon-Tiki*, was launched on March 23rd, 1995. With Heyerdahl's permission, Haslett christened his balsa raft *Illa Tiki*, the fire god. Thousands of local villagers turned out for the launch, and dozens of them insisted on climbing on board the raft as it took to the sea. Haslett feared disaster as the first wave rolled into the raft, but it merely lifted the balsa logs and drenched those on board.

Thirty-eight days later, from the top of the thirty-five foot mast, Haslett sighted Panama. In his first month at sea he had learned volumes about the coastal sailing abilities of a large balsa freighter. With the hope that he could modify the raft and improve its sailing characteristics, Haslett decided to anchor in Panama for a time.

It was while anchored in Panama that Haslett discovered that the nine balsa logs that made up the *Illa Tiki* were hopelessly infested with the wood boring *teredo navalis*. The raft had been in salt water for more than eighty days. The buoyancy of the raft had been so compromised that only modern pesticides would stop the attack. Haslett knew he had to reject such a course, for it would mean the end of his archaeological experiment and the beginning of a mere adventure, as well as a severe environmental hazard for the harbor in which the raft was anchored.

Crushed and dispirited, Haslett had no choice but to beach the *Illa Tiki* permanently on the shores of the Bay of Ciruelos in Panama. He realized that he had fallen into a familiar trap of mistaking adventure for science. If he was to make a real contribution to understanding prehistoric navigation, he would have to start all over. "On the *Illa Tiki* expedition we backed into the science," said Haslett, "because we started out merely to copy the adventure of *Kon-Tiki*."

> "The voyage to Hawai'i had been my primary goal. Now I realized that if we were serious about testing a prehistoric balsa raft we were going to have to sail it from point A to point B and test a scientifically grounded hypothesis. What we learned from *Illa Tiki* was that everything we knew about balsa rafts was wrong." (Haslett, personal communication, August 27, 2000)

Haslett knew he had the resources to launch another expedition, but he also knew that he had to do more than merely repeat earlier balsa raft drift expeditions. To do so, he felt, would only serve to remove the balsa raft from its native waters—the coastal seas of Ecuador and Columbia, and the ancient trade route all the way to Mexico. It was these coastal waters he now determined to explore.

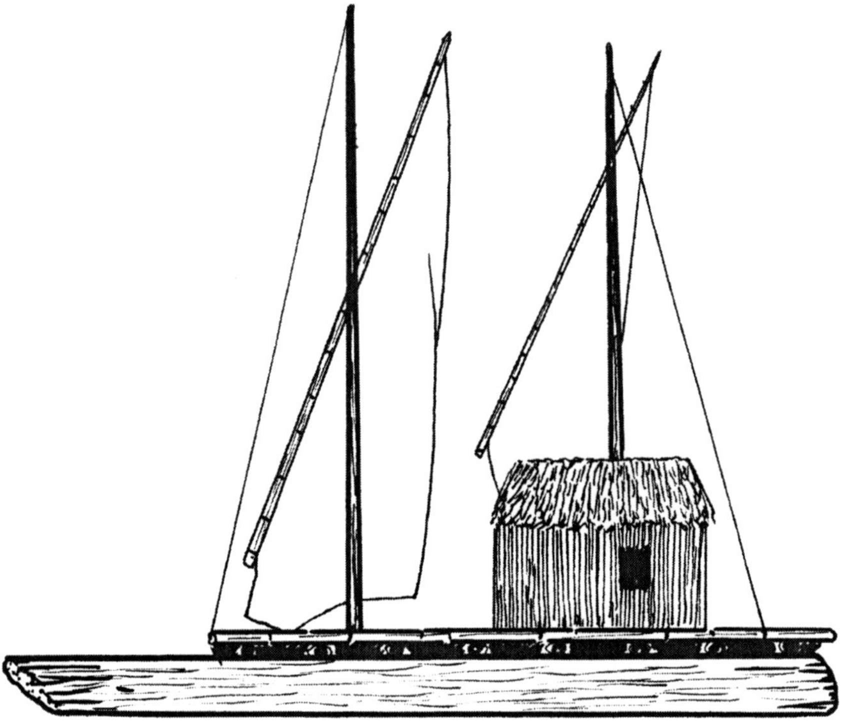

Fig. 16. Haslett's *La Manteña-Huancavillca*

La Manteña (1998)

By the fall of 1996, John Haslett had regrouped from the destruction of the *Illa Tiki* balsa raft off the coast of Panama a year earlier. With the fiftieth anniversary of the *Kon-Tiki* expedition only months away, Haslett wanted to both commemorate Heyerdahl's voyage and continue his own research in prehistoric Manteño coastal navigation by organizing his second balsa raft expedition from Ecuador. With good luck, his second raft would be ready to sail from Salango, Ecuador in August, 1997, a half century to the day after *Kon-Tiki* landed at Raroia.

To honor the ancient mariners whose technology he was recreating, Haslett decided to name his raft *La Manteña-Huancavillca* after both the Manteño culture and its immediate neighbor to the south. A local Salango shipwright, Maestro Enrique Guillen, once again oversaw the construction of the raft. And as with the *Illa Tiki*, nine balsa logs were lashed together with one and a quarter inch hemp rope. The longest log, which the locals dubbed "The Pope," was nearly sixty-one feet long. On top of these logs Haslett laid eleven crossbeams of heavy cocobolo wood. Forty, six-inch thick, bamboo logs lashed to the cocobolo formed the deck, atop which was a bamboo cabin topped with a roof of thatch palm fronds.

Haslett used another wood, guayacan, for the ten-foot long, sixteen inch wide guara centerboards, and two thirty foot pinuelo logs to form forward and aft masts. The fore and aft rigging was a radical departure from the square sail mounted on a bipod mast on the *Illa Tiki*. It was another indication that the new balsa raft would attempt to be an inshore sailing, not an offshore drifting, vessel. This meant new sails as well, and the expedition's archaeologist, Cameron McPherson Smith, spent a week with a Guayaquil tent-maker supervising the sewing of two triangular lateen sails. Gone was the familiar square sail carried by *Kon-Tiki* and so many other balsa rafts. With this new design, Haslett believed he had produced the truest copy of a prehistoric Manteño balsa freighter ever built in modern times.

A large part of that authenticity derived from the interpretation of the logbooks of early Spanish explorers by Haslett and Smith. They were convinced that descriptions of early native rafts as having "sails in the same fashion as our own" meant not square sails, but rather the kinds of triangular lateen sails required for coastal sailing. All previous replica raft builders had automatically assumed that such references meant that pre-Incan rafts were driven by the kinds of square sails the Spanish would have employed for deepwater downwind sailing.

To try and prevent *La Manteña-Huancavillca*'s destruction by *teredo navalis*, Haslett spent much of his time after *Illa Tiki* consulting with specialists on the marine mollusk such as Ruth Turner at Harvard University. On the first expedition, the balsa logs had been cut and then shipped by truck to Salango, unlike the spring of 1947 when Heyerdahl had floated his balsa logs down river from the interior. William Willis and Vital Alsar had also floated their logs down the freshwater rivers of Ecuador, and Haslett speculated that immersion in fresh water might have offered a kind of protection to the logs once they reached the salt water home of the teredo. Perhaps the logs of the first expedition, having missed any freshwater immersion, had not been 'inoculated' against the invertebrate.

On this second expedition, Haslett planned to sail *La Manteña-Huancavillca* up into coastal estuaries and river mouths whenever possible. In theory, such periodic natural shifts into freshwater environments would destroy any saltwater parasites clinging to the raft. The river excursions would also test the inshore navigability of the raft, while putting the crew in contact with upriver tribes that might have been part of the prehistoric trade network of the Manteño. Since Haslett also believed that another method for defeating the teredo could have been the periodic removal from the water and drying of the raft itself, the crew could also conduct surveys for potential prehistoric balsa raft landing sites.

Like Tim Severin, who experimented with natural lacquer as a way to guard his bamboo raft *Hsu Fu* against wood boring mollusks, Haslett also experimented with possible natural coatings to try to fend off the teredo. There was tar, found in naturally occurring tar pits and seeps along the Pacific coast and a plausible prehistoric anti-fouling paint. There was also the juice of the barbasco fruit, which was suggested to Haslett by local knowledge. This juice, whose active ingredient is called rotenone, was used to stun and catch fish, and one suggested method was to hang bags of basbasco underneath the raft, creating clouds of rotenone to frighten away any lurking teredo.

Experiments with submerging different blocks of balsa wood in seawater, however, showed that the juice had no effect on preventing a teredo infestation. As a compromise, and insure that this second balsa was not destroyed, Haslett coated the seven middle logs of the raft with antifoliant paint, and the two outboard logs with a mixture of 20% rotenone and 80% water. If was difficult—if not impossible—to know which, if any, of these natural methods were known or used in prehistory. The experiments provided at least some data on their effectiveness or lack thereof.

Haslett also planned two potential alterations from his itinerary on the *Illa Tiki* expedition. If *La Manteña-Huancavillca* reached Mexico, Haslett considered returning to Ecuador, thereby demonstrating a plausible prehistoric trading round trip. Or Haslett could construct two rafts: one to reach Mexico and continue on to Hawai'i, and a second raft to sail from Panama to Ecuador to explore the return voyage from Central to South America. In either case, the voyage to Hawai'i was now relegated to a secondary adventure, without any theoretical baggage carried on board. It was a clear signal that Haslett had graduated from adventuring, and had begun a new career as a true scientific experimenter.

Delays caused by funding and other problems—such as recruiting two separate raft crews—forced Haslett to scale back his second expedition to a single raft. Completed in Salango in September, *La Manteña-Huancavillca* was sixty feet long and weighed more than twenty tons. On September 28, 1997, the raft was pushed into the sea by a horde of locals assisted by a bulldozer. For two weeks, *La Manteña-Huancavillca* lay at anchor in forty feet of water while Haslett and his crew provisioned the raft for the expedition. They were two weeks Haslett was soon to regret.

On October 12th, during the local annual Balsa Festival, as the crew prepared the raft to be towed to sea, more than seventy residents of Salango poured onto the raft and most of the day was involved in setting them ashore. A fishing boat towed the raft two miles to sea and set it adrift. The crew raised the new triangular sails and headed north. Under overcast skies, Haslett tried to stay about thirty miles offshore as he navigated the raft along the coast.

Just twelve days after leaving Salango, Haslett sent two crewmembers overboard to check on the condition of the balsa logs. He was stunned to discover that teredo had already infested *La Manteña-Huancavillca.* The outboard rotenone-treated logs were severely infected, and the antifoliant paint on the main logs had flaked away in several places, allowing colonies of teredo to take hold of the raft. A balsa raft seemingly possessed acres of surface area, and short of immersing the logs in a vat of DDT, there were bound to be unprotected spots where teredo could enter. Once a single teredo bored into the wood, it multiplied and consumed the raft from the inside out.

The infestations forced Haslett to drive the raft ashore. Adverse currents kept the raft away from Panama, so *La Manteña-Huancavillca* was steered toward Colombia, to a place called La Playa de la Muerte, the appropriately named beach of the dead. Haslett was stranded, mystified as to why the relentless teredo seemed to single out his raft after a coastal voyage of 700 miles. Thor Heyerdahl, Willis Willis, Vital Alsar, none seemed to have had to deal with the destruction of their balsa rafts by teredos. *Kon-Tiki* had survived 4,300 miles of open-ocean and is now on permanent display in Oslo, Norway. Willis' *Seven Little Sisters* crossed 6,800 miles of ocean and was brought ashore in Samoa and put on display. *La Balsa* crossed the entire Pacific, more than 8,000 miles, and is now on display in Alsar's hometown in Spain.

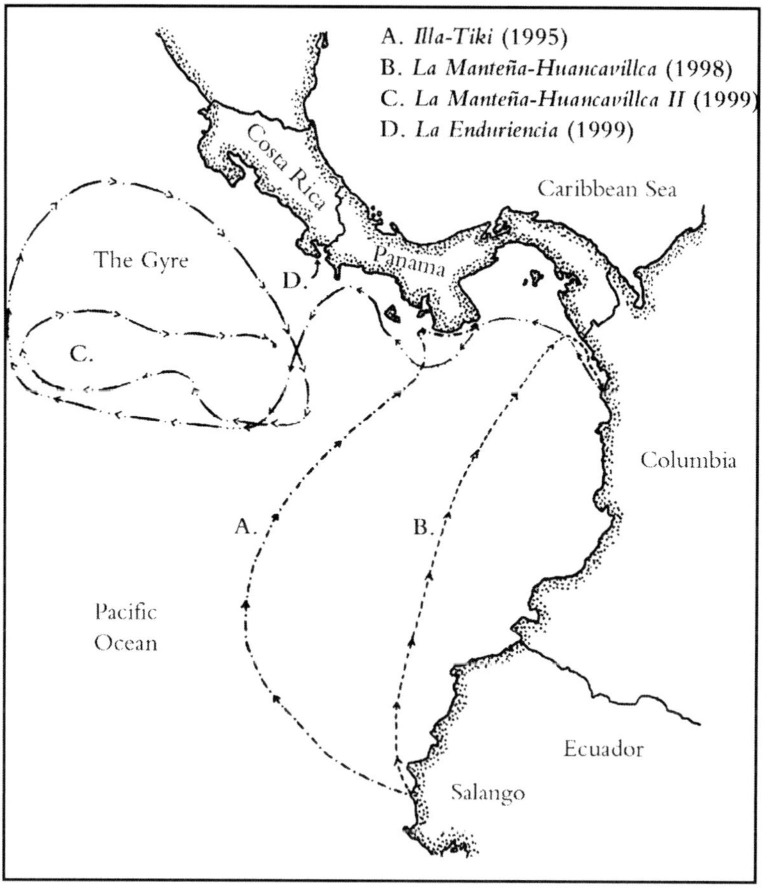

Fig. 17. Routes of Haslett's three balsa raft expeditions

As Haslett sat on the lonely Columbia beach, question after question filled his mind. Was it the very coastal nature of the voyage of *La Manteña-Huancavillca* that led to its infestation? Or the prolonged ride at anchor in Salango after the launch, a two-week period that could have allowed *teredo navalis* to get a deadly grip on the raft? How did the earlier balsa raft expeditions manage to avoid this menace? What of the prehistoric mariners he was trying to study? Was it possible that they never had to deal with the teredo? Could it be that Europeans unknowingly introduced the teredo to the Pacific? Burrowed into the wooden hulls of Spanish caravels, did this 'alien species' spread from Acapulco to the

Philippines, from Jakarta to Hawai'i, attached to the hulls of Dutch, British, or French exploration, trading, or military fleets? Or was it something more systematic and cultural, like regular switching of balsa logs, or charring the logs to keep the voracious teredo at bay?

Crewmembers began to leave the expedition and stray back to their homes, leaving Haslett with no good alternatives. Refusing to give in, Haslett replaced the balsa logs, and soaked these new logs with two coats of primer, three layers of natural brea, or tar, then a coat of white cement, and two more coats of tar. He replaced four departed crewmembers with two Columbians eager to join the adventure. The new, smaller crew now spoke primarily in Spanish. By late December 1998, the refitted *La Manteña-Huancavillca* II was ready to head to sea again, aiming for the west coast of Mexico. Ahead lay a treacherous stretch of Columbian coastline, one frequented by pirates known to board boats, kill their crews, and make off with the cargo.

Soon after the raft headed north at about a half a knot on January 1, 1999, it was caught in a pattern of circular ocean currents, known as a 'gyre,' off the coasts of Panama and Costa Rica. *La Manteña-Huancavillca* II found itself unable to break out. Irregular winds failed to fill the sails, and a near-hurricane and several smaller storms battered both the raft and its crew. Like Vital Alsar in 1966-67, weeks passed with little progress. Then, as the raft wound yet again around the gyre, Haslett discovered that his teredo nemesis had reappeared through seven layers of tar. As the raft slowly began to sink, the crew noticed a balsa log, of all things, floating by, and jammed it under the raft to increase its buoyancy.

After fifty days of aimless drifting, the main deck went awash. A brief east wind gave Haslett a temporary hope of maneuvering out of the gyre, but the wind also brought a heavy storm, washing equipment overboard and canting the mainmast. When the storm subsided, Haslett cut away the bamboo cabin, and threw overboard anything that could be spared and was biodegradable. With the raft lightened by nearly a ton, the crew had a notion to make a run for Hawai'i. But the gyre refused to give the raft up.

With the raft steadily sinking after another two weeks in the gyre, with a total absence of wind to push the raft, Haslett had reached the end of his tether. He had smoked cigars all his life, and one afternoon on the raft he could not understand why his face was surrounded by cigar smoke, something that had never happened before. It suddenly struck him that the air was so perfectly still that not even cigar smoke dissipated. He called a halt to the experiment. The crew was rescued by a cutter of the Costa Rican Coast Guard, but not before the captain of the cutter ordered Haslett to chop the drifting hazard apart and let the balsa logs float away separately.

Despite all his preparations and experiments, Haslett had been defeated a third time by a small marine invertebrate. It could be argued that prehistoric mariners knew how to defeat these pests. Perhaps they took a different route—further offshore to Cocos Island, for example, the island around which the dreaded gyre circulates—or sailed in a different season. Or perhaps the ocean currents and wind patterns were simply different a thousand years ago. But even more than the frustrating teredo infestations, the voyage of *La Manteña-Huancavillca* II had shown exactly how dangerous, lengthy, and potentially endless the trade route from Ecuador to Mexico could be.

La Enduriencia (1999)

Bjarni Grimolfsson's ship was blown into the Greenland Sea. They found themselves in waters infested with maggots, and before they knew it the ship was riddled under them and had begun to sink.

They discussed what to do. They had one ship's-boat which had been treated with tar made from seal-blubber; it is said that shell-maggots cannot penetrate timber which has been so treated.

—from the *Saga of Erik the Red*, ca. 1250

When the gyre forced John Haslett to abandon the sinking wreck of *La Manteña-Huancavillca II*, he salvaged what he could from the raft. In a titanic demonstration of will, once ashore at the small Costa Rican coastal village of Golfito, Haslett applied for and received government permission to harvest local balsa trees in order to build yet another balsa raft. He was now determined to sail this, his fourth balsa raft, directly to Hawai'i, teredos be damned.

Haslett built his new raft shorter and wider, to cope with storms the anticipated en route to Honolulu. Anticipating as well the wind at his back, Haslett stepped the familiar bipod mast to lift a square sail. As a nod toward his own personal hero, he named the new raft *La Enduriencia*, after Sir Ernest Shackleton's famous expedition ship that was trapped and crushed in Antarctica in 1915. Unfortunately, despite Haslett's Shackletonian leadership, *La Enduriencia* was doomed to a quick and crushing end.

A local powerboat towed the raft to sea, in hopes of avoiding a repeat trip into the gyre that had trapped *La Manteña-Huancavillca* II for more than fifty days. But the same tow pulled the raft straight into an offshore storm. Within a few hours, as wind and wave pulled *La Enduriencia* inexorably nearer to a rocky shoreline, the crew was forced to call for assistance. Haslett hoped that the Costa Rican Coast Guard would again come to his aid, with a tow into deeper waters. But the nearest patrol boat was forty miles distant, and by now the bipod mast had canted over, making the raft unmanageable. With the Coast Guard rescue still miles away, a wave picked up the raft and surfed it onto the rocks.

Even in the midst of such a colossal defeat, Haslett had lit the path ahead for future experimental balsa rafts. He had demonstrated conclusively that balsa logs do not become inevitably waterlogged and sink, as some anthropologists had long believed. Balsa logs themselves are incapable of absorbing enough water to make them heavier than the surrounding salt water. The logs will not fill with water unless they have been infested and the wood physically eaten away. What destroyed the buoyancy of a balsa raft was the teredo, pure and simple.

Moreover, by using over sixty different balsa logs, both male and female, cut down in all seasons, across three different countries, during every phase of the moon, with sap, without sap, Haslett believed he had put to rest many of the native myths concerning how and when balsa needs to be harvested. He felt that the key issue was the weight of the logs. For all practical purposes, a light balsa raft will float indefinitely. Therefore, prehistoric raft navigators must have possessed a system of beaching their rafts, both to dry the logs and retard the growth of teredo worms.

Far from demonstrating the weakness of the raft, Haslett's experience with the teredo led him to believe that the balsa raft was the only maritime solution for prehistoric coastal traders. If a teredo eats only a small hole in a traditional hulled vessel, it will sink out from under its crew. But a balsa raft can have as much as seventy-five percent of its hull eaten away, and it will still remain afloat and keep its crew alive.

Haslett could only watch as wave after wave smashed his fourth balsa raft to pieces at Golfito. Within a few minutes, the sails were torn apart and the yard snapped. There was nothing to do but accept the fact that the elements had cruelly conspired to destroy his expedition utterly and completely. One could only wonder whether prehistoric mariners had shown equal determination in the face of such adversity. Or whether, alone, exhausted, and shipwrecked on some unplanned island, they too had accepted their fate, and fell onto the beach, leaving the sea at their backs once and for all, to take their chances with a new life ashore.

Chapter 9

The Norwegians Return to the Pacific

Tangaroa (2006)

On a bright early summer morning in June, 2004, Torgeir Sæverud Higraff and I sat on the wall near Akerhus Fortress that overlooks Oslo Harbor. He had read Sea Drift in its original English, and asked if we could meet when I came to Oslo to study Norwegian in the summer of 2004. Torgeir wanted to recreate a balsa raft expedition in the Pacific, and name his raft *Tangaroa* after the Polynesian god of the sea.

Torgeir was in many ways typical of Norwegian men I had met in ten years of traveling in and around Norway, from Oslo to Svalbard. He lived an outdoors, adventurous life. Tall, strong and rugged, he had been a river guide in western Norway, but now sought much wider horizons. He was impatient not to advance further into adulthood without a major expeditionary accomplishment behind his name.

We met near the National Theatre station and walked towards the waterfront. Torgeir was restless, almost agitated. I had seen this many times before in expedition leaders. It was the natural behavior of an explorer consumed by expedition planning, financial uncertainties, crew recruitment issues, and the ever-present scrutiny of colleagues, friends and family.

Apart from his wife, he had few people to turn to who understood his dream of building a balsa raft and following Thor

Heyerdahl across the Pacific. When, a few weeks later, we visited the hytte of his in-laws near Kragerø, he made me promise not to discuss raft voyaging or his intention to build a balsa raft in Ecuador in the spring of 2005. Like Thor Heyerdahl 60 years earlier, he was certain of his wife's support, but not at all certain about her parents. He had not told them about his plans, and was entirely unsure how they would react.

But his biggest concern, like John Haslett a decade earlier, was how to frame the science he viewed as underpinning the expedition. Sea Drift had shown him that there were two distinct types of raft expedition leaders: those, like Heyerdahl, who encountered adventure in their pursuit of science; and those who had little or no use for science and instead sought adventure for adventure's sake. Torgeir was a Norwegian after the model of Thor Heyerdahl, and he clearly saw little use for a new balsa raft expedition if it could not contribute to science.

Looking over the modern and historic sailing ships in the harbor, we spoke at length of the idea of recreating a prehistoric ocean. Balsa rafts sailed originally on a prehistoric sea, one that is difficult to imagine today. Current archaeological research suggests that the original inhabitants of South America may have arrived from Melanesia or Australia, and predated a later Asian migration by thousands if not tens of thousands of years. If true, then experimental archaeologists working to recreate prehistoric rafts need to account for wind and weather and current and island patterns at several different points in the oceanic past.

We continued our discussion during a weekend in Kragerø, which started with a side trip to Thor Heyerdahl's hometown of Larvik. There, we boarded a small Norwegian sjekte on a crystal clear and calm harbor, and made our way from the dock at Stavern to the small island where Heyerdahl played as a boy. The sjekte then chugged along at a slow pace to the narrow channel that splits the rock islet of Kråkøya. At high tide, we had just enough water to allow the sjekte to navigate the channel with only a few centimeters on either side of the boat.

We discussed what the Pacific Ocean looked like during the height of the last ice age, when sea levels were hundreds of feet below what they are today. How many islands now underwater would have appeared above the surface? How would this radically different geography have affected one's ability to navigate across the Pacific? Torgeir promised to keep this prehistoric ocean in his imagination as he attempted to cross the modern Pacific.

Later that year, after the disastrous tsunami that devastated coastal countries around the Indian Ocean and shifted corporate and personal donations to relief efforts, Torgeir was forced to postpone the *Tangaroa* expedition until 2006. But there was not the same tone of discouragement that usually accompanies such an announcement. Instead, Torgeir wrote that preparations for a voyage in 2006 were already well underway. He had identified the right kinds of balsa trees during a trip to Ecuador. The Peruvian Navy had agreed to host the building of the raft at the naval base at Callao from where *Kon-Tiki* launched in 1947. And, in Stavern, a half-scale replica of the balsa raft made several test voyages around the harbor, within sight of the granite monument dedicated to Thor Heyerdahl's four expeditions on prehistoric vessels. Heyerdahl's own grandson, Olav, had agreed to join the expedition as a crew member.

By January 2006, Torgeir and his crew had journeyed from the freezing winter in Oslo to the warmest part of Ecuador to harvest the balsa logs for the expedition. Using a drilling tool specially made for Scandinavian forestry, Olav Heyerdahl drilled the first hole in an enormous balsa tree identified the previous year and named "Torgeir." After the logs' samples were declared strong enough for the raft, Torgeir and his crew felled a dozen logs and created two massive river rafts from them. As Heyerdahl had done in 1947, they then began a five-day drift down the Quevedo River to Guayaquil.

Heavy rains allowed the balsa rafts to meander the 100 kilometers to Guayaquil at about three kilometers an hour. The rafts glided past small villages and huts peopled by river fishermen. Torgeir watched as the river people picked their breakfasts directly from the fruit trees along the banks, then discarded their waste into the same waters that sustained their lives. The crew halted and slept on the rafts at night, anxious to avoid any low-hanging power lines strung across the river.

In Guayaquil, the logs were transported to a warehouse where the materials needed for the mast, the guara, the cross beams and the bamboo were waiting. The journey down the river also revealed that two of the 12 balsa logs were not fit for an ocean crossing, and had to be replaced before the raft could be built. After arranging for the supplies of wood to be shipped from Ecuador to Peru, the crew made its way south to Callao in mid-February. It would take nearly a month to get the logs and the wood for the crossbeams and the guara from Ecuador to Peru.

In Peru, the team found a building site at the Peruvian Navy shipyard. While awaiting shipment of the containers from Ecuador, they lived in rooms at the Naval Academy and began to splice the three kilometers of rope needed to join the balsa logs together. After weeks of delays, the logs finally arrived in Callao, along with four Norwegian carpentry students from Larvik who would help with the cutting and shaping of the balsa into a transoceanic raft. More time was lost waiting for the tools necessary to build the raft to clear Peruvian customs.

As the tools arrived, so did the *Tangaroa* crew begin to come together: four Norwegians, a Swede and, just a few days prior to sailing, a Peruvian. Besides Torgeir, now 32 years old and the expedition leader and historian, and Olav Heyerdahl, 28, there was the raft's captain and doctor, Bjarne Krekvik, 52, a sailor with experience on board Viking ship replicas. Øivin Lauten, 54, the oldest member of the crew, was an experienced coastal sailor and was in charge of the construction of the raft. Anders Berg, a 42-

year-old Swede who had worked on documentaries with Thor Heyerdahl on Easter Island, became *Tangaroa*'s filmmaker. And Roberto Sala Rey, 44, a Peruvian sailor, became *Tangaroa*'s navigator.

Time was now of the essence; launch date had been fixed for April 28, exactly 59 years since *Kon-Tiki* was launched from the same spot. Important visitors had made special arrangements to attend the launch ceremonies; no delay was possible. Torgeir and his crew now had little more than four weeks to take a dozen balsa logs and create a raft that could survive thousands of kilometers of open ocean.

Just as the crossbeams finally arrived to join to the raft, the Norwegian carpentry students had to return to Larvik. Torgeir drafted several workers from the local shipyard, and together they began the frenetic work of lashing the balsa logs together, laying on the crossbeams, and then adding on the guara, the bamboo deck, the hut and the masts, all before the rigging the raft. Launch date was now only three weeks away.

Even in the face of such a challenge, the *Tangaroa* crew managed to find time to visit an orphanage and a school for poor children run by a Norwegian foundation and supported by Norwegian students who give the income of one day's work every year to the project. Torgeir brought the school several copies of *Kon-Tiki*, and with the support of a secondary school in Larvik named after Thor Heyerdahl, both the orphanage and the school were connected to the Internet so they could follow *Tangaroa* across the Pacific through newsletters in Spanish.

The *Tangaroa* crew met the launch deadline with two days to spare. As Torgeir wrote in his blog from Callao:

> On Friday the 28th of April the expedition will start their long journey towards Polynesia, on the very same spot and the same date as Thor Heyerdahl launched the world famous *Kon-Tiki* raft in 1947. At that time, the journey took 101 days, ending at the Raroia reef. Today's heavily equipped balsa raft

> is bigger, has a three times as large sail and will, as opposed to *Kon-Tiki*, be maneuverable by the same techniques as applied by the prehistoric South-American seafarers. The expedition will, supported by the University of Bergen, Norway, look for molecular pollution and document any changes in the sea water since 1947.

As the raft drifted away from the coast of Peru, the crew began to sense that *Tangaroa* indeed outperformed *Kon-Tiki* on several levels. When the sails were finally raised on the morning of the second day at sea, the crew felt the immediate acceleration of the raft. After experimenting with the correct placement of the guaras, which were too long and had to be cut down to maintain the raft's sailing balance, they settled on a course of 270°, which took them more or less directly towards the west. Compensating for the northward drift of the Humboldt Current, the plan was to stay on a north-westerly course but in a flatter, more southerly arc than that taken by *Kon-Tiki*.

With the winds coming on the beam, the raft plowed westwards at 2.5 knots. Gradually the crew learned to fine-tune the placement of the guara to maintain their course. As Torgeir wrote on May 1st, "If the wind drops, we just raise or lower one of the guaras at the stern to stay on our course. This happens so rarely that the helmsman can easily keep in complete control of the raft in between the passing fishing vessels and the reading of a good novel from his place in the hammock."

When the winds backed from the west a few days later, however, the crew attempted to turn the raft towards the east. The stress on the bamboo spar that held the main sail caused it to snap. Reinforced with hard wood brought from Ecuador, the sail was soon back up, the wind shifted around to the east, and the raft making 3 knots towards Polynesia. By *syttende de mai*, as the crew decorated the raft with Norwegian flags, it was clear that *Tangaroa* was on a considerably flatter course than *Kon-Tiki*. The raft did not come close to the islands of the Galapagos that had doomed earlier raft expeditions. The raft was in fact on a nearly straight-line course from Peru to Raroia, the Tuamotu island where *Kon-Tiki* landed in 1947.

After the crew raised the flag, Captain Bjarne began the holiday celebration with a speech. After singing the National anthem, the crew sat down to a breakfast of risengrynsgrøt and fresh baked bread. In the spirit of the day, a Norwegian flag was even placed underneath the raft so the fish could share in the Norwegian's patriotism. Five days later, already one-third of the way to Raroia, the crew calculated that they were 350 nautical miles ahead of where *Kon-Tiki* was at this same point in its drift. By using the guara to maintain a Great Circle course straight to Polynesia, the raft was moving at the prehistoric equivalent of light speed.

The first gale met up with the raft more than a month out from Peru. The heavier winds pushed *Tangaroa* even faster, at times approaching 4.5 to 5 knots. On the 29th of May, the raft made 79 nautical miles. A few days later, as Torgeir hauled in the same species of bizarre snake mackerel (Gempylus serpens) that Thor Heyerdahl had caught in 1947 and in approximately the same place, the raft broke its own record and made 80 nautical miles in a day. As Øyvin Lauten wrote: "Waves of 5 metres snapping at us from behind, sending buckets of water up on deck. The wind is blowing straight in from behind, and we just sit and watch the raft sailing by herself. Even if the weather is quite heavy, the sun is shining, and we're rushing towards the west!" On only one day did the winds die down completely, and then the crew took the chance to have a swim without worrying about the raft leaving them astern.

After two months at sea, *Tangaroa* was only 500 nautical miles from Raroia. They were less than a week away, a rate that would put them into the Tuamotu Archipelago in just 65 days, more than a month faster than the 101 days *Kon-Tiki* required for the same voyage. The balsa logs were only an inch or two lower in the water than when the raft left Peru, none of the hundreds of sisal rope knots had loosened, and the Seglloftet sails made by a Norwegian family company were still driving the raft at more than 3 knots. As Torgeir wrote in his blog:

"After two months at sea we are still friends. Fact is that we are even better friends now than in Callao. Here it's easy to discuss topics, to get a full understanding of the view of our crewmates. In Callao, we were in a dire straight the last month. Never time to talk things through. And March stressed us to the edge, because of the overwhelming logistical challenge. On the raft, so far, we all agree that we share favorable conditions for making a good society: Every single member of this society has agreed upon the rules, and these rules have been made for the safety onboard. Today is our 60th day at sea. Just "another day at the office" is a common joke in the morning, before a swim in the biggest clear water pool in the world."

Five days later, the crew maneuvered the raft around Raroia and its dangerous reefs, and brought the raft to anchor. Towed to a more favorable anchorage on the west side of the island, a reception that had been planned for months got underway. The vice-president of French Polynesia, accompanied by Miss World 2004, landed at Raroia's newly constructed airstrip, and helped the best outrigger canoe paddlers on the island welcome the *Tangaroa* crew with flower leis and a song performed by local children.

The hospitality on Raroia was the same as that which had greeted Heyerdahl and his crew. The differences lay in the intrusions of the modern world. In addition to the new airstrip, a small brick building Heyerdahl had visited in 1947 was showing the World Cup on a large-screen satellite television.

From Raroia, the crew again headed westwards, away from the Tuamotu Archipelago and toward Tahiti and the Society Islands. They sailed between the islands of the Tuamotus, trying to find the fastest route to Tahiti. In an area of shifting currents, dangerous reefs and heavier traffic, the crew counted on the winds remaining strong enough to keep them on course.

When *Tangaroa* reached Raiatea, another welcome ceremony was held, this time with 1,000 people at the temple platform (marae) ruins of Taputapuatea. *Tangaroa* had sailed more than 4,500 nautical miles, nearly halfway across the Pacific, in little more than two months. Towed from Raiatea, the raft was welcomed in Papeete, Tahiti, after 93 days of sailing from Peru. They had drifted farther and faster than *Kon-Tiki*, landed at Polynesia's ancient cultural center of Raiatea, and demonstrated that an improved sail rig and active use of the guara centerboards allowed the raft the steer a direct course from Peru to Polynesia.

Just two weeks after his return to Norway, I met again with Torgeir. This time we took the ferry across Oslo harbor on a hot late August day and sat on the shores of the small island of Gressholmen, discussing his adventure. We talked about a plan is to create a Thor Heyerdahl Centre in Larvik, with the *Tangaroa* raft the enduring symbol of Heyerdahl's life and legacy. And we talked of his desire to get away from the hustle of Oslo and move to the western coast of Norway, into the heart of the Viking homeland, write his book about the adventure, and anticipate parenthood with his pregnant wife.

Torgeir was tired, and admitted that had he known the amount of work it would take to build and sail the raft, he would have thought twice about the expedition. But he had accomplished his great adventure, and the sense of agitation I saw in him two years earlier was gone. He could now proceed into middle age with a sense of himself. It was all anyone could ask out of drifting across an ocean on a balsa raft.

Chapter 10

Other Experiments After Kon-Tiki: Science

Feathered Serpent
La Balsa
La Tres Balsas
Uru
Chimok
Titi
Mata-Rangi
Mata-Rangi II
Abora
Kota-Mama I, II, III

Abstract

While many explorers took their inspiration from the balsa raft *Kon-Tiki*, Thor Heyerdahl's American Indians in the Pacific (1952) compiled a sizeable mountain of evidence that pointed toward the possibility that pre-Incan mariners in reed boats had once attempted to cross the Pacific Ocean from South America to the islands of Polynesia. Not until thirty years later, however, with his *Tigris* expeditions, did Heyerdahl learn exactly how long a reed ship, constructed from properly harvested reeds, could stay afloat. Beginning with Gene Savoy, a new generation of expeditions sought to use this knowledge to test the limits of the reed boat on experimental voyages from South America into the Pacific.

Feathered Serpent (1969)

> The upper jungles, the length of Peru east of the Andes, contained evidence of ancient civilizations [and] many incorporated into the architecture the concept of a great personage. This historic figure was known to the old Peruvians as Viracocha, among other names. According to legend, he sailed away to the north following a period of persecution. I believe he emerged in Mexico as Quetzalcoatl, the *Feathered Serpent*.
>
> —Gene Savoy
> *On the Trail of the Feathered Serpent* (1974)

> Nothing concerning the first European encounter with America has puzzled the rest of the world more than the Aztec claim that the Spaniards were not the first white and bearded people to have reached them from across the Atlantic Ocean.
>
> —Thor Heyerdahl
> *Early Man and the Ocean* (1979)

Like Thor Heyerdahl, American explorer Gene Savoy had an intense interest in the prehistoric civilizations of Peru. But where Heyerdahl fixed his interest on the peoples of the ancient Peruvian coast, Savoy ventured across the Andes and deep into Amazon rain forests to locate and excavate cities very few even believed existed. And where Heyerdahl speculated that prehistoric mariners could have drifted from Peru to Polynesia and built *Kon-Tiki* to prove it, Savoy's archaeological research led him to similar conclusions but in opposite directions.

Rather than reaching across the Pacific from South America, Savoy was convinced that the Andean empire of the Incas and the Mexican empire of the Aztecs had had regular contact along the western coasts of South and Central America. In 1961, Savoy had "observed a petroglyph etched on a rock in the Jequetepeque Valley of northern Peru. It was a Mexican hieroglyph for gold" (Savoy 1974, 17). Observing other similarities, it seemed clear to Savoy that a regular prehistoric trade had existed between the two culture areas, and that that trade was conducted over the water.

Savoy's archaeological research along the steamy eastern slopes of the Andes led him to associate the mythical Peruvian figure Viracocha with the Aztec figure of the Feathered Serpent Quetzalcoatl. At the time of the *Kon-Tiki* expedition, Heyerdahl imagined a defeated Viracocha at the head of a fleet of balsa rafts escaping across the Pacific to Easter Island. Savoy interpreted the evidence from both Peru and Central America to mean that Viracocha had fled from Peru northwards toward the valley of the Mexica, to emerge as the mythical Feathered Serpent of the Aztec realm.

For Savoy, it was clear that "trading fleets [had] operated between Panama and Peru. They went where they liked and were capable of extended voyages, but kept within fifteen or twenty miles of the shore." (ibid, 19) To demonstrate his belief, Savoy set out in the spring of 1969 to build a double-hulled ship built of totora reed and sail it from Peru to Mexico. Savoy named his proposed vessel Kuviqu; his own contraction of the names of Viracocha, Quetzalcoatl, and the Maya figure Kukulcan. The *Kuviqu*, or *Feathered Serpent*, as it was more often referred to, took shape at a dockyard in the small fishing village of Huanchaco. Local fishermen still used floats made from reeds for their daily fishing trips. Savoy noted some forty of these twelve-foot-long reed bundle caballitos de totora on the beach.

Savoy's research had led him to the intriguing possibility—tested again by several reed boat expeditions in the 1990s—that pre-Incan reed boats had either two hulls, like a modern catamaran, or two sterns, like a kind of arrow made from reeds. He had ordered four large pre-made reed bundles from Lake Titicaca, but when they arrived at the coast he found that they had rotted. So he turned to the local fishermen to save the expedition. When Savoy showed one a sketch of what he had in mind, the man replied that he could build Savoy a catamaran of two oversized caballitos, thirty-six feet long.

Each of the twin hulls of Savoy's "slow-moving drift-type vessel" (ibid, 49), was formed from four separate totora floats. The reeds were bound around thirty-foot lengths of bamboo, which, according to the fishermen, would stiffen the floats should the totora begin to weaken and threaten to break apart. The amount of reed necessary to construct the raft forced Savoy to scour the coastal plains. Eventually he secured a large supply by purchasing it, appropriately enough, from the reed-growing reservoirs near the ancient centers of Chan Chan and Moche. When finished, each float weighed only about 250 pounds. A layer of bamboo reinforcements was lashed under each float, as a kind of protective fender for the reed, and bamboo formed the deck and cabin.

The final fitting out of the *Feathered Serpent* took place at a naval dockyard at Salaverry, eighteen miles south of Huanchaco. The twin hulls were attached to the decking, and the cabin built and the double mast stepped. Savoy on the one hand refused to employ nylon rope, as it would detract from the authenticity he was trying to achieve. Yet he decided against native woods for his masts in favor of Oregon pine "in honor of my home state" (ibid, 65). Dacron sails, modern navigation instruments, an inexplicable citizen's band radio, and primus stoves were put onboard the catamaran. Two serpent-headed guara centerboards were dropped between the hulls to complete the rig. As Savoy himself pointed out, all of the modern devices and navigational aids made the voyage less authentic, but then again, "ancient Mochica and Chimú sailors didn't … have to contend with large ships of steel hundreds of feet long, traveling at fast speeds" (ibid, 77).

In the end, one had to admit that no true replication of an ancient voyage was possible. One worked with modern knowledge, employed present-day humans to build the vessel, and used as source material prehistoric designs painted on pottery, artistic representations always subject to differing interpretations. Savoy's raft builders presented him with another modern-day problem. They insisted that the ancients had coated their totora reed boats with varnish to keep marine borers from eating the bundles. They therefore showed up with several gallons of varnish and a spray gun and applied a coat of varnish to the reeds and the bamboo. A local bank donated ceramic coin banks made to look like pre-Incan pottery.

On April 15, 1969, after dumping an overload of supplies that caused the reed boat to ride too low in the water, the *Feathered Serpent* was towed to sea and let go. The raft was taken by the Peruvian Current and carried northwards. Sluggishly it labored back toward shore, as the crew fought to get control of the large steering oar. With first nightfall and two seasick crewmen, Savoy let go the oar and allowed the raft to drift until morning.

The next day, the *Feathered Serpent* made its first, unscheduled landfall, to retrieve more wood to widen the sweep oar. A fishing boat towed the raft in too fast, as the Chilean naval vessel had done to de Bisschop's *Tahiti-Nui* a decade earlier and several hundred miles south, and the raft began to break apart. Figureheads were torn loose. Totora reed broke off in large chunks. Only by cutting the tow with a machete was the raft saved from destruction.

Savoy retreated ashore to a hot shower, while totora reinforcements reached the raft from Huanchaco. Journalists swarmed over the apparently mortally wounded raft, while Savoy began to calculate his next move. If the raft managed to get to sea again, he would be required to keep well offshore to avoid being thrown back on the coast. But as he approached the border with

Ecuador, if he drifted too far offshore, the Humboldt Current would push the raft toward the Galapagos, and threaten the whole hypothesis behind the expedition.

The next day was a desperate one for Savoy. "Word had spread that we were landlubbers dependent on towboats" (ibid, 96-97). The sweep oar was widened and extended to give the raft more control, and new figureheads attached to the twin prows. Savoy determined to sail the raft out of its temporary harbor, but again the raft exhibited a mind of its own. The lateen sail drove the raft so fast to the southwest that Savoy had it hauled in and stowed, never to be used again. The crew raised the square sail, and drifted slowly westward before the wind. The chase boat from the CBS television network left the raft on April 18th, and *Feathered Serpent* was thereafter on its own.

After nine days at sea and two ashore, the raft drifted into the port of Talara, some 350 miles north of its starting point at Salaverry. There Savoy was accused of trying to transport antiquities out of Peru, and the raft was detained briefly. The raft was then sailed out of Peru and into Ecuadorian waters. On May 6, 1969, the raft was towed into Manta, the port where Savoy believed that Viracocha had fled South America to emerge in Central America as Quetzalcoatl. Once again, a port call almost destroyed the expedition.

When Savoy requested that the raft be lifted out of the water so he could inspect the wear on the totora reed, the *Feathered Serpent* was accidentally dropped from fifteen feet in the air. The shock of impact broke the vessel's back, canted the deckhouse, and collapsed the double mast. In no position to repair the damage at this stage of the expedition, Savoy transformed what had been a sailing reed boat into a true drifting cargo raft.

The expedition returned to sea on May 14th, drifting northwards toward the equator. Stuck in blistering heat, Savoy longed for the cool Andean highlands. A storm sprang up and lasted for over a week, during which time the crew assumed they were being pushed north of Columbia, across the Gulf of Panama, and on to Panama itself. Savoy's navigator confidently took a few sun sights, and assured Savoy that the progress of the raft northward had been steady and strong. By the end of May, the navigator announced that the raft had emerged from the storm to a position off the coast of Costa Rica, exceeding even Savoy's wildest predictions.

In fact, when the raft drifted past a local fisherman on June 6th, he told the crew that they were afloat off the coast of Cabo Marzo, Columbia. The navigator's positions had been off by more than 500 miles. Later that day, the raft was driven onto a shoal by contrary currents, and Savoy was forced to break out his emergency raft. The crew discovered that the emergency inflatable boat was large enough for two men, not four as Savoy had ordered but had not bothered to confirm. Unable to put the crew into the escape raft, Savoy put his cameras on board instead, and all were subsequently waterlogged, destroying his visual record of the expedition.

Five days later, a local canoe towed the raft five miles offshore, and the *Feathered Serpent* began a meandering drift northward in search of the current that would sweep it across the Gulf of Panama. The current picked up the raft on June 14th, and propelled the battered reed raft northwards. Three days later, Savoy anchored his reed boat for the last time, in Panama Bay. Savoy himself went off for a month of exploring Mayan ruins, and when he returned, he discovered that a storm had blown his raft to sea. He never saw his *Feathered Serpent* again.

La Balsa (1970)

In the spring of 1970, just a few months after Gene Savoy's *Feathered Serpent* drifted into oblivion and Thor Heyerdahl's first reed boat broke apart just short of Barbados, Spanish explorer Vital Alsar built his second balsa wood raft. Alsar already had one balsa raft expedition to his credit, the ill-fated Pacifica, caught in 1966-67 north of the Galapagos islands for more than one hundred days in the dreaded Equatorial Counter Current. Alsar's ambition for his second raft, which he christened *La Balsa*, was nothing less than a drift across the entire Pacific, in the wake of William Willis' metal raft expedition of 1963-64, and more than twice as far as Heyerdahl's *Kon-Tiki* in 1947.

Alsar took as his prehistoric cultural model the Huancavilca Indians of Ecuador, coastal mariners who twenty years later would provide similar inspiration for John Haslett's three balsa raft expeditions. As a young man in the Spanish Foreign Legion in the early 1950s, Alsar had been transfixed by a Spanish edition of *Kon-Tiki*, and resolved to drift across an ocean on a recreated Huancavilca balsa raft himself. Later, studying to be a language professor, he read the French edition of *Kon-Tiki* as well. He also cited the work of the Argentine anthropologist Juan Moricz in supporting his belief that the prehistoric Ecuadorian mariners were no strangers to the islands of the Pacific.

> [Moricz] points out that the Huancavilcas thought of the ocean as "a forest of rivers," with predictable currents to and from the Polynesian islands. They also knew about "friendly and unfriendly" winds and the use of astronomy in navigation. Commenting on the presence of South American cocoa trees, quecha kuka, in the far-off Mexican highlands, Moricz concludes that the Ecuadorian natives had sailed all the way

> to Mexico long before the conquest of Montezuma by Hernán Cortés. (Alsar 1973, 23)

La Balsa was built without the use of modern materials, all the way down to a specially designed balsa wood privy, which dangled over the port side of the raft providing instant waste disposal. The raft was towed from the harbor at Guayaquil, Ecuador, at two o'clock in the morning of May 29, 1970, and set adrift when the towboat had reached the far side of the Gulf of Guayaquil. Among the four crewmen was once again the Frenchman Marc Modena, chef and crewman from previous drift expeditions on *L'Egare II* across the Atlantic and on Pacifica.

By the end of the first week at sea, the raft had reached the eastern edge of the Humboldt Current. Once he had a clear day, Alsar hoisted the mainsail especially painted for him by Salvador Dali. The explorer had met Dali in Montreal, and several days later, forgetting the time difference, had called the artist when it was three in the morning in Spain to beg him to paint the sail for the raft. Moreover, Alsar explained that he had no money to pay for it. "Como tienes cojones, hombre," Dali had responded (ibid, 63), but the great artist in fact did paint the sail. When finished, it showed a small balsa raft atop what for all appearances looked like a setting sun, or like an extremely large pair of cojones, depending on which art school you attended.

In a similar fashion, Alsar kept his crew entertained during the long days adrift with a superb wit. It was a skill Heyerdahl himself felt most vitally necessary on a long expedition. Humor was absolutely essential to keep a crew calm in a crisis, or bring a sullen loner out of his shell, or defuse a potential fight in the cramped quarters of a slow-moving raft. Alsar had likewise prepared for the possibility of severe injuries or shark bites by observing surgical procedures firsthand, but he knew that nothing could keep his hands totally still on a constantly rocking raft if such a crisis situation arose.

Like all other drift voyagers, the crew of *La Balsa* came to see their raft as a floating reef, populated by a whole range of marine invertebrates, dolphins, sharks, and other fish during the day, and lit up by phosphorescent plankton at night. At one point Modena, the French chef, created a casserole from rice, spices, and a few barnacles. The crew was duly impressed. "Had he used a bit of saffron," Alsar wrote, "it might have tasted like a poor man's paella." (ibid, 95) They were surrounded by an irony, an entire ocean where swimming, because of the ever-present sharks, was largely forbidden. The decks of the raft were likewise treacherous, not the least reason being an army of ants that had infested the wood before the expedition began, and which now swarmed over everything.

After hitting a stretch of blistering heat and light or absent winds in early July, the middle of that month brought a new wind from the east, and the raft began scudding along at more than five knots an hour. On July 14th, *La Balsa* advanced 132 miles westward. On July 30th, the raft passed 142°05'W, the longitude of Raroia island, where *Kon-Tiki* had landed after 101 days in 1947. *La Balsa* passed a thousand miles north of the island on its 62nd day at sea. It was here where Alsar thought his real challenge would begin. He wanted to show that an Incan or Huancavilcan balsa raft could navigate not just the open ocean drift from South America to Polynesia, but the far more dangerous passage between the islands and reefs of the western Pacific all the way to Australia.

At this point, the condition of the balsa logs assumed paramount importance. Heyerdahl had noted the decreasing buoyancy of his balsa raft at this point, just before the end of his voyage, and William Willis had been likewise concerned that his balsa logs were losing buoyancy and slowly sinking during the drift of the *Seven Little Sisters*. Alsar cut into each of his balsa logs and each seemed to be riding as high and dry as when he left Ecuador. He was convinced that this was because his logs were female rather than male, and had been harvested with less sap in them, and so therefore lighter. This is the opposite argument made by Heyerdahl, who thought it was that very sap which acted to keep out the salt water and preserve the buoyancy of *Kon-Tiki*'s balsa logs.

In early August, Alsar's routine transmission to a ham radio operator in Mexico became garbled, causing the false impression that the crew were starving and in danger of dying. A U.S. research vessel dispatched from Pearl Harbor caught up with the raft in western Polynesia. The unexpected appearance of the American ship provided Alsar and his crew with a dinner of steak and fries, along with boxes of supplies when they returned to the raft. Yet no sooner had the ship maneuvered out of sight than most of the supplies were pitched guiltily overboard, as the drifters sought to return to a recreated state of prehistoric purity.

By the end of August, *La Balsa* had drifted two-thirds of the way from South America to Australia. Upon inspection, the balsa logs seemed to be retaining their buoyancy, even after more than 6,000 miles at sea. Winds and currents propelled the raft southwestwards towards Samoa at nearly 130 miles a day. Off the coast of Samoa, Alsar's raft was taken in tow by one of the same islanders who had helped William Willis ashore as *Age Unlimited* approached the islands in 1963. Alsar related how he himself had met Willis at the Explorers Club in New York in 1965, before the *Pacifica* drift expedition. Willis, then seventy-two, at first discouraged Alsar, then warmed to his obvious determination. Like the coincidental confluence on Tahiti in 1956 of Eric de Bisschop, Thor Heyerdahl, Norman Baker, Bengt Danielsson, and Ben Finney, the meeting of Alsar and Willis in New York represented once and future transoceanic expeditions involving tens of thousands of miles.

Fierce storms dogged the raft between Samoa and Fiji, and by early October *La Balsa* was snaking through a minefield of reefs and shoals north of New Caledonia. Beyond these lay the Great Barrier Reef, which the raft cleared on October 28, 1970, followed by Saumarez Reef a day later. The raft now drifted lazily toward the Australia. Early in the morning of November 5, 1970, the raft picked up the lights of the Australian coast. Alsar noted that his balsa logs rode only about an inch lower in the water than when the raft departed Guayaquil more than five months earlier and 8,500 miles away. Given the condition of the logs, he thought that *La Balsa* would have been entirely capable of making a round trip all the way back to South America. Alsar arranged for the raft to be shipped all the way to his hometown of Santandar in Spain, and then his crew dispersed back to their respective homelands.

La Tres Balsas (1973)

Almost incredibly, Alsar was true to his word, and returned for yet another round of balsa raft drifts, in the summer of 1973. This time, he built a fleet of three balsa rafts, *La Aztlan*, *La Guayaquil*, and *La Mooloolaba*, and drifted away from Guayaquil harbor on May 27, 1973. The twelve crewmembers consisted of Alsar and, yet again, Marc Modena, along with three Americans, three Canadians, two Chileans, a Mexican, and an Ecuadorian. After a voyage of 179 days and more than 9,000 nautical miles, this prehistoric fleet arrived in Bellina, Australia, on November 21, 1973. With this most ambitious of balsa expeditions, Alsar had by now drifted on board balsa rafts for nearly 500 days, and more than the distance of the circumference of the earth at the Equator. Marc Modena, through his three voyages with Alsar as well as the transatlantic expedition of Henri Beaudout, had now drifted more than 25,500 miles on wooden rafts, which almost certainly ranks him first in the history—and very likely the prehistory!—of transoceanic rafting.

One of the three rafts, *La Aztlan*, was donated to the Maritime Museum on Norton Street in Bellina. Appropriately, it is featured at the museum alongside the *Florrie*, a river tug built in 1880. For no balsa raft had yet made it to sea under its own capabilities, and without the assistance of a modern tugboat.

Uru (1988)

In 1988, a Spanish explorer named Kitín Muñoz, who has referred to himself as the "spiritual son of Thor Heyerdahl" (Reuters, April 27, 1997) set out to build a reed boat and drift all the way to Tahiti on it. The *Uru* was similar in size to *Ra II*, but constructed of totora reeds instead of papyrus. It was built by the same reed boat builder, Paulino Esteban, who along with other Aymara Indians, had built Heyerdahl's successful transatlantic reed vessel. On June 29th, 1988, Muñoz set out from Lima, Peru, and in seven weeks, crossed the eastern Pacific to a landing in the Marquesas. Muñoz had every intention of continuing his voyage by continuing on to Tahiti, but a severe storm set the reed boat adrift until its rescue by a Tahitian fishing vessel in October. Muñoz and his crew were brought ashore, and the expedition ended, on October 16th, 1988. He had shown that a reed vessel could in fact survive the nearly two month voyage from South America to the nearest islands of Polynesia.

Chimok (1991)

By the late-1980s and into the mid-1990s, with the approach of the fiftieth anniversary of the *Kon-Tiki* expedition, the focus of explorer-adventurers attempting to drift from place to place shifted back to the South American coasts of Peru and Ecuador. With few exceptions, explorers during this period took Heyerdahl's vast thesis and spun from it a series of intensely interesting expeditions. These experimental voyages were of a largely different character from those of Willis or Genovés, or even Alsar and Severin. With the exception of the overreaching reed boat voyages of Kitín Muñoz, they explored relatively minute hypothetical questions that Heyerdahl's enormous theory generated. In so doing, they reached a kind of final refinement of the voyage of the *Kon-Tiki*, and an ultimate acceptance of the global ocean as an archaeological sea.

As Heyerdahl himself was intensely involved in the excavations of the coastal pyramid complex of Túcume in Peru beginning in 1987, several explorers approached him for advice and assistance in getting their drift expeditions underway. One of these was the ill-fated voyage of the reed boat *Chimok*. A German film crew sought to sail a reed ship from a port in northern Peru to the Galapagos islands. They would thereby provide an experimental linkage between pre-Incan cultures of the coast of South America and the pre-Incan pottery shards found in the Galapagos by Heyerdahl and Arne Skjølsvold during an expedition to the archipelago in January, 1953.

Paulino Estaban, who had helped to build Heyerdahl's *Ra II* and *Tigris*, and Muñoz' *Uru*, arrived in the Peruvian fishing village of Pimentel to construct the *Chimok* from totora reeds. Heyerdahl suggested to the film crew that they try to build a double stern reed boat, one that Heyerdahl suggested was of Moche design. As Heyerdahl recounted from his own local experiment in building and testing a small double stern reed boat, the Moche design possessed distinct advantages:

> Apart from better stability and increased space for a larger bamboo deck, it also solved the awkward problem of getting back on board if one dived into the sea or fell overboard. A reed-ship is like a big barrel floating high on the water, making it exceedingly difficult to get on board again unless someone throws out a rope or a ladder. But with a split stern it is possible to swim into the fork and climb up with ease. (Heyerdahl 1995, 223)

The *Chimok*, with its dragon heads arising from the double stern, was launched from the beach at Pimentel in front of a large crowd, with Heyerdahl doing the honors of christening the huge vessel. Almost immediately, the Germans, along with Paulino and a local Peruvian fisherman, sailed into a storm, but not one of nature's creation. As the reed ship floated perfectly atop the waves, the crew received word that tensions along Peru's border with Ecuador had escalated almost to the point of war. The Ecuadorian government insisted that the *Chimok* sail into the port of Guayaquil and obtain proper clearance papers before attempting to sail for the Ecuadorian-controlled Galapagos.

In film of the voyage, a narrator relates that the Ecuadorian government threatened to send a warship after the reed boat if it did not turn around. The crew speculated that the Ecuadorian government resented the *Chimok* expedition specifically because Heyerdahl harvested the balsa logs for *Kon-Tiki* in Ecuador yet sailed from Peru. South American politics aside, it is difficult to imagine a national government holding a grudge against a forty-year-old archaeological experiment.

As Heyerdahl's archaeological teams were unearthing evidence of a direct connection between the birdman cult on Easter Island and the Lambayeque culture dated to roughly 1,100 AD in pre-Incan Peru, the reed boat *Chimok* was being abandoned in Peruvian waters and allowed to drift on alone into the Humboldt Current. Heyerdahl's teams unearthed adobe friezes that seemed to show birdmen dancing atop reed boats or balsa rafts fitted with apparent bamboo cabins, while surging waves surround the vessel. A miniature silver paddle and birdman ornament were also excavated, both with striking similarities to paddles and birdmen found on Easter Island.

Heyerdahl speculated that a disastrous El Niño event around 1,100 C.E. had scattered the Lambayeque to sea, possibly to fetch up on Easter Island. Or perhaps the Lambayeque leader himself was simply trying to avoid the fate of a karakas, the hereditary rulers of prehistoric Peru. Karakas acted as conduits between earth and the dictates of heaven and, as Michael Moseley writes, when things went wrong in the natural environment, the karakas paid with their lives. "[One] powerful potentate in Lambayeque, was put to death when a great El Niño devastated his homeland." (Moseley 1992, 52)

No satellite tracking of the *Chimok* was apparently possible before it was abandoned, so it is not known for how long nor how far the reeds drifted before they became waterlogged and sank. But the expedition was a powerful reminder that an archaeological experiment is not conducted in a sterile laboratory, where variables can be tightly controlled. An archaeological experiment is conducted in the modern world, in real-world geographies, with definite implied cultural consequences for living peoples. Those people might not necessarily want an outsider sailing away from them or toward them with some pet hypothesis about who their ancestors might have been or where they might have come from. Archaeological experiments with prehistoric rafts are not conducted on prehistoric seas.

Titi (1993)

The Catari family, who live along the shores of Lake Titicaca, have a long-standing familiarity with explorers and anthropologists seeking to test a favorite hypothesis on a drift expedition. When a scientist or explorer shows up in Bolivia looking to test the maritime capabilities of a reed boat, they usually leave Lake Titicaca in possession of a reed bundle boat constructed by the Cataris. They have rightfully become famous around the globe for their reed boats. There is even a Catari family museum in Huatajata, Bolivia, with some of their finer models on display.

In 1993, when Bolivian scientists hypothesized that the ancient natives of Tiahuanaco used reed boats to transport the rock they used in monumental architecture at such sites as the Gateway of the Sun, they came to the Cataris to talk about how to build such a ship. The family patriarch himself, Maximo Catari, decided to help construct and sail the vessel "to recuperate my ancestral heritage" (Bolivian Times, 18-25 November 1994). This ancestral heritage looked back across two millennia to the same Tiahuanaco culture often cited as having carried their stoneworking techniques to Easter Island, several thousand miles out into the Pacific.

The reed boat *Titi* was build over the course of three months in 1993 by Erik Catari, another family member. Catari decorated the ship with three figures from Aymara/Inca legend. On the prow was the cat-like Mauk'a Pacha, the Aymara Indian god of the underworld. The sail was painted with the principal figure of the Gateway of the Sun, Akapacha, and the masthead carried the celestial god Alajja Pacha.

When finished, *Titi* was fourteen meters long and three and a half wide, and had cost some $8,000 to build. In October, sailing in the extreme cold of the nights to take advantage of more reliable winds, the vessel sailed completely around Lake Titicaca in twenty-eight days. The reed boat carried six anthropologists, who to the surprise of nearly all observers did not overly tax the vessel's carrying capacity of eight tons.

Mata-Rangi (1997)

On Easter Island in 1997, aided by the Explorer's Club, the United Nations Educational, Scientific, and Cultural Organization (UNESCO), the Spanish government, and a watch company, Kitín Muñoz began building the largest prehistoric vessel ever constructed by modern humans. At a reported cost of over one million dollars, the reed ship extended 131 feet and was constructed entirely of totora reeds. The reed vessel was christened *Mata Rangi*, a contraction of the Polynesian "Mata ki te Rangi," or "eyes that look at the sky," a reference to the outward looking statues on Easter Island. With this new vessel, more than twice the length of *Uru*, Muñoz planned an even larger expedition—a voyage around the world. Stops were planned in Tahiti, Australia, and Japan, as the reed ship made its way across the Pacific.

Muñoz' eleven-man crew spanned the Pacific from Bolivia to New Zealand, and included a Hawai'ian, two Tahitians, two Maori, two Aymaras, three Rapanui or Easter Islanders, along with Muñoz himself. "The name of our expedition ... unites all (of) these people who are so different," Munoz told an interviewer. "Mata-Rangi means 'eyes of paradise,' because what it hopes to do is look at our planet, which is our paradise" (Lovler 1996). The chance to study the navigational capabilities of ancient civilizations would be accomplished with the help—or in spite of —the modern navigation instruments put on board, which included a radio, satellite telephone, radar and GPS.

To build the enormous ship, Muñoz and his team of Aymara indians used reeds from the crater lake inside the extinct Easter Island volcano of Maunga Terevaka. At least one of the Aymara builders had his doubts as to the buoyancy of the Easter Island reeds, which he thought weaker and more water-absorbent than the related reeds that grow at Lake Titicaca (Te Rapa Nui 11 (3): 4). The ship's oversized design was derived from the form of a ship scratched into an Easter Island moai statue. "It is a classic boat based on old designs and prototypes," related Muñoz (Lovler 1996). This particular statue, named No. 263, and the ship carved into its chest, was excavated by Norwegian archaeologist Arne

Skjølsvold, during Thor Heyerdahl's archaeological expedition to the island in the 1950s. Skjølsvold (pers. comm, March 14, 2000) has come to believe that the ship on the statue probably represents a European sailing vessel, one carved into the statue after it was erected.

Muñoz was justifiably proud that his team used only its hands and bundles of reed to build the boat, without any mechanical assistance. But then the hull waited for ten months before its final fitting out. It was then that machine labor—in the form of two boats and a crane—was forced into action to push the ship into the sea when manual labor failed to budge the massive ship off the beach. When the expedition was at last ready to leave in late April, 1997, the launch of the seventy-ton craft was delayed by bad weather. There was fear that the ten months on the beach might have rotted many of the reeds.

Mata Rangi left Easter Island on May 6th, 1997. Muñoz had hoped to sail the massive boat west to remote Pitcairn Island, landing place of the *Bounty* mutineers, and then on to Tahiti. But just two days from Easter Island, the Spaniard discovered two stowaways on board. The two had been turned down when they asked to join the expedition when it was learned they had been accused of rape and robbery. Muñoz now had thirteen crewmembers and, as one might predict, his luck began to run out.

Radio contact with the ship was lost on May 12th. Four days later, when contact was reestablished, the world learned that *Mata Rangi* was only one hundred and forty miles northwest of Easter Island, when they had expected to travel at least 600 miles in that time. Muñoz ditched Pitcairn as a destination and announced he would make directly for Mangareva. By May 17th, eleven days out from Easter Island, a satellite telephone communication received in French Polynesia alerted authorities that the reed boat was trying to stay on course with a broken mast. Two hundred miles north of *Mata Rangi*, the Chilean training ship *Esmeralda* moved to intercept the primitive vessel and offer any assistance. When the naval vessel found the reed ship, it was lying partly submerged, with ocean water pouring in the gap created when the mast fell.

According to one news account: "The crew of the Esmeralda noted that some of the 13 men on board the *Mata Rangi* were "prisoners of fear""(Lovler 1996), yet the expedition continued on. Eighteen days out from Easter Island, on May 24th, *Mata Rangi* split in two and the crew was forced to abandon ship. The men tied rubber lifeboats to the sinking hulk, and hung on to await rescue. When a sailboat found the *Mata Rangi*, the reed ship was only one hundred and eighty-five miles northwest of Easter Island. The Eyes of Paradise had looked out over the Pacific for less than three weeks.

Even in failure, however, the expedition provided valuable insights into reed boat construction and navigation. Heyerdahl had learned that to form an effective reed ship the reeds themselves must be harvested in August to preserve their buoyancy. By remaining for so long on the beach at Easter Island, the reeds that made up *Mata Rangi* likely had more than enough time to become brittle, or rot, and in either case lose their flexibility and their natural buoyancy.

The very size of the vessel, as well, seemed to contradict the lessons learned by Heyerdahl and his crews in the design, construction, and sailing of the *Ra* ships and the *Tigris*. Reed ships bigger than twenty meters seemed to work themselves apart and lose buoyancy quickly. The lack of a rope wrapped around the length of the reed bundles, a so-called circumferential rope, allowed the reed bundles to work apart longitudinally. Heyerdahl employed such a rope on all three of his reed ships, and not one of them worked apart longitudinally.

Finally, the experience of the *Mata Rangi* seems to offer support to Skjølsvold's feeling that the ship carved into the No. 263 moai is indeed a European vessel, and not a prehistoric reed boat. The No. 263 carving has three masts, and appears square-rigged. If *Mata Rangi* is any indication, a three-masted reed ship is an extremely dangerous ocean-going vessel, one unlikely to have made a successful voyage from the shores of the South America to Easter Island. Whatever came within sighting distance of those prehistoric stone-carvers, looking out to sea from the slopes of the extinct volcanoes of Easter Island, it was more probably a three-masted sailing ship from Europe.

Mata Rangi II (1999)

By the end of the 1990s, the port of Arica in northern Chile began to challenge the great Peruvian naval port at Callao as the favored point of embarkation of experimental raft voyages. Kitín Muñoz returned to South America early in 1999, intent on demonstrating that the *Mata Rangi* disaster was an aberration and that the 1988 voyage of the *Uru* was more typical of the transpacific capabilities of the reed boat. Supported again by the Swiss watchmaker Breitling, and the United Nations Educational, Scientific and Cultural Organization (UNESCO), which named the Spaniard an ambassador for indigenous peoples, Muñoz again saw as his mission the cause of furthering science and fraternity.

Now forty years old, Muñoz constructed the *Mata-Rangi II* on the beach at Arica with totora reeds trucked to the coast from Lake Titicaca. Construction began in the fall of 1998. The new vessel was smaller, but not enormously so, than the original *Mata Rangi*. When finished, the reed boat stretched thirty meters in length, weighed some fifty tons, and was fitted with the same three-masted rig as its predecessor. Once again, Muñoz had chosen to interpret the ship carved onto the chest of Easter Island moai statue No. 263 as a prehistoric three-masted reed ship, rather than an Easter Islander's vision of a passing European square-rigger. On the bow of the new boat was a figurehead representing the birdman cult of Easter Island. The mainsail was painted with similar birdman representations.

Like *Mata Rangi I*, the new reed boat was fitted with modern navigational instruments including a satellite uplink and a global positioning system. Solar panels ran power to the ship's mess area, where wicker baskets were filled with rice, oatmeal, cans of carrots, potatoes, spinach, and kidney beans. Muñoz proposed a route that would take the reed boat from Arica all the way to Micronesia, an archipelago north of New Guinea and almost clear across the Pacific, a voyage of more than 8,000 miles over four months. If he was successful, Muñoz anticipated that eventually he could reach the Asian mainland, or at least Taiwan or Japan. It was an ambitious programme for an explorer whose previous reed boat broke up less than 200 miles from land.

In February 1999, a local Arica tugboat towed *Mata Rangi II* to sea. Muñoz' international crew varied according to which newspaper account one read. One said that it consisted of three Easter Islanders, a Peruvian, a Bolivian, two Japanese, a Tahitian and an Arican; another had it manned by two Easter Islanders, two Peruvians, two Aymara Indians from Bolivia, one Japanese and one Tahitian. Muñoz himself told a reporter that he would pass the time on the long passage to Japan by studying and letting his mind wander.

His peace of mind did not last long. The vessel began to break apart as ropes became worn and began to give 'way. As the reed boat drifted toward the Marquesas it quickly became a reed raft. The crew was forced to shorten the massive raft, cutting off the bow, and two of the three masts were lost overboard. From its proud start at Arica, the reed boat took on the shape of a floating haystack, its crew clinging to it like a group of tenacious barnacles. Though Muñoz promised to continue his projected circumnavigation of the world by reed boat, as the Marquesas approached it became clear that it would not be on board the *Mata Rangi II*.

Even in failure, Muñoz had shown, for the second time, that a Lake Titicaca reed boat could in fact make the 5,000-mile journey from South America to the Marquesas. Through his three expeditions, he had offered experimental evidence that reeds from Lake Titicaca seemed superior as boat-building materials to those reeds that grow in the crater lakes of Easter Island. Again he had shown that an enormous three-masted reed ship was not a likely candidate as the flagship of a prehistoric fleet. And he had again offered experimental evidence that the ship carved into Easter

Island's statue No. 263 is more likely a square-rigged European vessel, and not a prehistoric three-masted totora reed ship.

Abora (1999)

In 1990, a German high school teacher named Dominique Görlitz formed a group of students and interested adults to investigate prehistoric navigation and construct modern copies of prehistoric watercraft. The group succeeded in building three small reed boats, the *Dilmun* I, II, and III, which exhibited an ability to sail as much as 75° against the winds of the Baltic Sea.

Encouraged by this success, Görlitz and his group of volunteers in 1999 decided to vastly enlarge the scope of their experiment. He sought to extend Heyerdahl's earlier *Ra* experiments, both in sailing his reed boat against the wind, and in closing the gap from Egypt to Morocco left open by the earlier *Ra* expeditions, which transported their reed boats overland from Egypt to the Atlantic.

This new experiment would make use of currents circulating counter-clockwise around the Mediterranean to transport the reed boat from Sardinia to Gibraltar, along much the same course drifted by Alain Bombard in his rubber Zodiac nearly half a century earlier. If one assumed that any vessel put into the ocean off the coast of Morocco would automatically be carried to the Americas, as the *Ra* expeditions demonstrated, then the *Abora* expedition would accomplish the really hard part of the experiment that *Ra I* and *Ra II* avoided by sailing directly from the port of Safi on the Atlantic. In many respects, this is one of the most difficult experimental questions of all—the study of how prehistoric mariners exited the Mediterranean.

Harvesting seventeen tons of reeds from a marsh near Leipzig in the fall of 1997, they transported these materials to the island of Sardinia in the summer of 1998, where they constructed an eleven-meter long reed boat that was christened *Abora*. The crew was proud that they could build such a craft with only manual labor, at least until they came to the middle bundle of reeds, which weighed over a ton and a half and had to be lifted into place with a crane.

On May 22, 1999, *Abora* departed from Sardinia and promptly sailed all the way to Italy. The reeds, possibly from the extended time between their harvest and the launch of the vessel, almost immediately became waterlogged and began to sink. Perhaps, the crew speculated, ancient Egyptians had not sailed to Gibraltar via Sardinia. In either case, Görlitz seemed far from discouraged. He immediately planned a new expedition, one that would start from the Balearic Islands further west—and from which Alain Bombard launched his adventurous and ultimately unsuccessful attempt to escape from the Mediterranean. From there, the *Abora* II would attempt to drift and sail its way beyond the Pillars of Hercules.

Kota-Mama I, II, III (1998-2000)

Colonel John Blashford-Snell, OBE, DSc (Hon), FRSGS, one of the most famous explorers of the twentieth century, served for thirty-seven years in the British Army, leading the first descent of the Blue Nile in Ethiopia in 1968 and the first expedition from Alaska to Cape Horn across the Darien Gap in 1972. He pioneered the use of inflatable boats for rapid descents of river gorges. Thereafter, through his Operations Drake and Raleigh, he led more than sixty expeditions involving thousands of children from the world over receiving their first experiences in field exploration.

Later, Blashford-Snell organized the *Kota-Mama* expeditions, intent on exploring possible maritime cultural connections between the Americas and Africa. "I am convinced there must have been more intercontinental voyaging in ancient times than history credits," the Colonel wrote (John Blashford-Snell, personal communication, 7 Aug. 2000).

Kota-Mama I, carried out in March and April of 1998, involved a flotilla of reed boats making their way from Lake Titicaca to

Lake Poopo along the 250-mile long Desaguadero River. The expedition located four archaeological sites and through an education programme linked students in Bolivia with students in England.

Kota-Mama II, a much more ambitious project, took place from July through October of 1999. Using traditional Lake Titicaca reed boats, Blashford-Snell and his teams voyaged 1,800 miles from the foothills of the Andes to the Atlantic Ocean at Buenos Aires. The experiment carried out the Colonel's trademark wildlife and archaeological surveys combined with educational outreach programmes. But perhaps its greatest contribution was the demonstration of a plausible route for the cultures of the high Andes to conduct trading expeditions with cultures both on and across the Atlantic.

The latter notion, that prehistoric reed boats could have crossed the Atlantic from west to east in a kind of reverse *Ra* expedition, is the central theme of planned *Kota-Mama III* and *IV* expeditions. Three different river routes were surveyed in the summer of 2000, exploring for plausible reed boat and balsa raft routes from the Andes to the sea. The third *Kota-Mama* expedition, "will involve archaeological quests and taking a nine metre traditional reed boat plus a balsa raft from the Andes to the Atlantic via the Rio Mapiri, Rio Beni, Rio Mamore, Rio Madeira, and the Amazon." (ibid.)

The Colonel pointed out that such an expedition, involving over 4,000 kilometers of river travel, included a stretch of 500 kilometers of sharp rapids. The chance to test both reed boats and balsa rafts in such waters will allow the explorers to evaluate whether or not interior cultures could have maintained large-scale water-borne communication with the rest of the prehistoric world. This is the truly unique aspect of the *Kota-Mama* expeditions—their attempt to use primitive reed and balsa rafts to link the interior of the South American continent with the sea. Given the importance of Gene Savoy's discoveries of major cities on the eastern, jungle areas of the old Incan empire, deep in the Amazon watershed, these expeditions may offer new insights into the possibility that these highly advanced peoples of the interior were not in any way insular or provincial.

Chapter 11

Other Experiments After Kon-Tiki: Pseudo-Science

Lehi I
Lehi II
Lehi III
Lehi IV
Acali

Abstract

After the sinking of the bamboo raft *Tahiti-Nui*, Eric de Bisschop intended to build a copy from the same materials for a return voyage to Tahiti. But with no bamboo available in Chile, where Bombard was rescued, he instead had a new raft built from buoyant Chilean cypress wood. There was no experimental reason for this choice. It was made simply because there was no other suitable wood available.

For this reason, despite the success of *Tahiti-Nui* as a drift experiment from Tahiti to the offshore waters of South America, the *Tahiti-Nui II* expedition can be seen as one of half a dozen raft expeditions based not on science but on absurd notions that were at best pseudo-scientific. Bisschop could at least claim a long and detailed interest in the problems of Polynesian diffusion. The other pseudo-scientific expeditions could claim no such distinction.

Lehi I (1954)

It was perhaps inevitable, given the notoriety gained by *Kon-Tiki*, and by Alain Bombard's raft trip across the Atlantic in search of the secrets of survival, that sooner or later during the 1950s someone would use a raft expedition to fight against communism. And that was less than the half of it. The Lehi raft expeditions, organized by a retired California shipyard owner named DeVere Baker, sought to drift across the Pacific not merely to defeat the atheistic collective, but to turn Thor Heyerdahl's thesis of a bearded white god bearing the gifts of civilization into the second coming of Jesus Christ.

Devere Baker was a devout Mormon and, after *Kon-Tiki*, an individual with a weakness for proposing raft voyages as the only true method to confirm various theological doctrines. In the same summer that William Willis used his balsa raft to voyage into his own inner spirit, Baker built a plywood raft in San Francisco in order to transform the politics of the entire population of the planet. According to Baker, between *The Book of Mormon* and *The Antiquities of Mexico*, there existed more than enough evidence that a "great, fair, highly cultured race of people [had] sailed from the Old World to the New World on 'caves of wood.'" (Baker 1959, vii).

Here in radically racist form was the idea of a light-skinned, bearded culture bearer bringing the concepts of civilization to dark-skinned savages around the world. The only problem, in Baker's view, was the dark-skinned savages refused to stay civilized. The bearded Semites from the Middle East—be it Christ enjoying a resurrection in prehistoric Mexico and/or Peru around 34 C.E., or the Israelite prophet Lehi bringing the gospels to America around 600 B.C.E.— were continuously dumped or duped by natives who refused to see the light.

> The people of one nation were white, delightsome (sic), cultured and peace-loving, living within the cities; the other people were fierce, warlike, and given to wild paintings of the body, weird firelight ceremonies and waging war upon their

> more peaceful brothers. Is it one wonder their bare bodies became darkened and their natures inflamed by the animal-flesh diet which was their mainstay [for it was these] dark-skinned people [who] finally killed, scattered and absorbed through intermarriage the white peaceful race who were their brothers, so that generations later only the brown-skinned Indian was left in the New World to greet the Spanish conquerors. (ibid, 18)

Baker cited the still-mysterious Olmec "Uncle Sam" stone relief from La Venta, and an equally inscrutable Olmec pottery figure from Tres Zapotes, to argue for the presence of stocking-capped Israelites in America. Conveniently ignoring the Aztecs and Incas, Baker saw the Spanish arriving in a New World peopled only by naked dim-witted heathens. Such widely scattered threads and staggering speculations were more than enough, for Baker, to verify the Second Coming of Christ. Once that claim was accepted, he felt, the existence of God would thereby be proven. With a literal God, atheistic communism would be unmasked as fundamentally flawed. Taken together, Baker could then lay the foundations for a new era of Christian peace amongst all the peoples of the world.

To glue together this house of cards and prove his theories to the world, Baker conceived the only logical experiment possible: a drift expedition to Hawai'i on board a small raft. He chose his destination, as he wrote, because the ancients had used the Japan Current to drift from Asia to the New World and, as everyone knew, a branch of the Japan Current bends toward Honolulu after it hits the west coast of America. Q.E.D.

> By chance these islands were first discovered by the inhabitants of North and South America as their sailing vessels were blown from the Japanese current off course... In the ancient records [sic] of these people, we read of a ship

> builder named Hagoth who excelled in his trade, and of how some of the ships of commerce set sail and were never heard of again. Without doubt a majority of these ships caught an offshore wind and were blown into a branch-off [sic] of the Japanese current, then drifted and were blown to the Hawai'ian Islands (ibid, 20-21).

Baker had also contrived to use his raft expedition to study survival at sea for the U.S. Navy, even to the point of convincing Jon Lindbergh, son of Charles, to join the trip and leave the dock without food or water. (Lindbergh later backed out when his participation was leaked prematurely and against his wishes to the media.)

Admittedly, it was one tall order of hypotheses for a small raft hammered out of Douglas fir plywood. So it was hardly a surprise that the *Lehi I* was abandoned just a week after its launch from San Francisco, barely out of sight of the Golden Gate Bridge. The Coast Guard rescued the five men on board, but refused to tow the raft in. The abandoned *Lehi I* washed ashore on Guadalupe Island, but Baker could not get there in time before the tide carried her back out to sea. The raft was last seen adrift in 1958, "on her way into the Humbolt [sic] Current, consistently averaging her forty miles a day—as did the ancients who came to, or left from, this continent hundreds of years ago." (ibid, 40) Once again, Q.E.D.

Lehi II (1955)

The second Lehi expedition fared little better than the first. According to Baker, a crewmember enamored of publicity set the second raft adrift without his permission during the last week of April, 1955. By the time Baker caught up with the errant raft, it was bobbing helplessly in a storm. Once again, just beyond the Golden Gate Bridge, the U.S. Coast Guard plucked the crew of *Lehi II* from the sea. And once again, the Coast Guard refused to tow the hulk to shore. So it was written, so it was done, that the second Lehi raft followed its unmanned predecessor along the route of the ancient culture bearers to Oahu.

Lehi III (1957)

Now more than eighty thousand dollars out of pocket, yet determined as ever, Baker scaled back the conception of his third raft. The *Lehi III* was little more than a sixteen-foot-long floating cabin surrounded with a rope walkway. Yet this third raft, launched from San Francisco in March, 1957, actually succeeding in bobbing drunkenly all the way down the west coast to Los Angeles. En route, Baker lost his lone crewman, who stormed ashore cold and dispirited at the pier at Kuikus. After three days alone, Baker himself drifted into the port of Avila where, exhausted by the long tortuous days at sea, he took on board two volunteers.

This crew of three not only completed the drift the rest of the way to Los Angeles—demonstrating no doubt a plausible route to that city for semitic wise men from the east (perhaps the Lower East Side)—but sold a film of the voyage to a television program called "I Search for Adventure." It was success enough to convince Baker that true redemption, the voyage to Hawai'i, was just around the corner. And indeed it was.

Lehi IV (1958)

Nineteen fifty-eight was a big year for Pacific raft voyages attempting to show who were the real Polynesians, and who the mere squatters. On July 14, 1958, as Eric de Bisschop and his raft crew were struggling with thirst, madness and, later, death on board the *Tahiti-Nui II* in the mid-Pacific, DeVere Baker launched *Lehi IV* from the fair city of Redondo Beach, California. With Lehis *I* and *II* still afloat somewhere between California and Jakarta, the competition for Roman numeral space on the ocean seemed almost to foreshadow the coming competition between the American Mariners and Surveyors and Soviet Lunas and Zonds in the ocean of space.

Baker constructed his fourth raft through the expedient of dropping, literally, his third raft on a bed of logs on the beach at Redondo. Baker was highly pleased that this feat was accomplished without the aid of "the mechanized might of a rented crane" (ibid, 76). Instead, all he required were a few ancient steel cables attached to the bumper of a primitive Lincoln, and 'man's determination to do the impossible.' The launch itself was delayed for a time when Baker, hurrying to his bare-chested appearance before television and newspaper cameras gathered for the event, mistakenly put both legs into the same leg of his swim trunks and nearly fell flat on his face.

The expedition proceeded down the coast and then offshore to Guadalupe Island, which seemed to exert a powerful and mysterious pull on the Lehi rafts. Given the sophistication of his concerns, however, it seems a bit bereft at this point for Baker to ignore the empirical evidence for prehistoric Israelite landings on Guadalupe. When it became clear the voyage was doomed to end just 250 miles from Redondo Beach, on the same shore where Lehi I stuck on the rocks, Baker was forced to start his ten-horsepower outboard in order to push the twelve-ton raft around the island.

Two of the crewmembers on board *Lehi IV* had voyaged with Baker already from Avila to Los Angeles on board *Lehi III*. The fourth crewmember, a Hawai'ian college student named Ed Kekaula, seems to have been brought on board to play the ukulele and represent the "child-like people" (ibid, 275) of Polynesia. When the raft, with its huge square sail painted with a copy of the Olmec "Uncle Sam" image, arrived off Maui on September 20, 1958, Baker was grateful that Kekaula could act as a translator for him as he sought to communicate with the exotic islanders. Baker seemed surprised to learn that a Hawai'ian who swam out to the raft could actually speak English. Once again, the Coast Guard arrived on the scene, offering to tow the *Lehi IV* into Kahului Harbor on the island of Maui, but a Japanese tuna boat beat them to it. The wise men from the east had arrived in Polynesia.

Once ashore, Baker squabbled with the Maui Historical Society over donating the *Lehi IV* to their museum. According to Baker, the museum asked for, then accepted, then rejected, his offer of the raft for a display. Baker then steamed into the Bishop Museum, where he felt he had uncovered information that the first people to land in Hawai'i were "predominantly Near Eastern" (ibid, 325). Kenneth Emory, dean of all Pacific archaeologists, was kind enough to receive him in his office in the museum. The Mormon Baker was profoundly shocked when Emory rejected his suggestion that, with the success of *Lehi IV*, the museum should now include "both theories of migration in the museum information (the migration from Central America as well as the one from the direction of the Gilbert Islands)." (ibid)

Thus chastened that a mere academic would dismiss his proven theory without first exposing it "to public attention for the public to draw its own conclusions" (ibid.), Baker retreated to the saner confines of the Waikiki Biltmore. "Each morning in our rooms was a fresh pineapple with the inside sliced lengthwise and replaced to look as though it had just left the bush. This was only a beginning!" (ibid, 331). Mu-mus, hula lessons for his visiting daughters, nightly floor shows, shopping for Hawai'ian shirts, invitations to "a party for Billy Eckstine, the melodious Negro singer" (ibid, 334), it was more than enough tourist junk culture to restore the faith of even the most tainted of pioneering monotheistic raft captains.

DeVere Baker wound up his Hawai'ian triumph with lectures at the Latter Day Saints temple in Laie. Thus fortified, he announced plans for his greatest Lehi expedition yet. It was to be a long and complicated drift from the shores of the Red Sea all the way to the Central American birthplace of the ancient Olmecs (not forgetting a stop in Redondo Beach) via the Indian Ocean and the Japan Current. He would drift along the entire presumed trajectory of the bearded white men from Israel. The expedition, however, never seems to have materialized, though Baker in 1969 claimed to have found a meteorite in Australia that proved the existence of God. There was also something about a dog that joined the Navy and a large-breasted spirit woman who taught culture to heathen scientists, but after awhile it all seemed to run together.

It would be fitting if some future archaeologist, toiling away on the shores of Guadalupe Island, discovered the final proof that Jesus really did walk in the New World. Then adventurous DeVere Baker could have the last laugh on wise old Dr. Emory and the brochure-printers of the Bishop Museum in Honolulu.

Acali (1973)

> I am aware that a minimal raft, adrift in a huge sea, would constitute an ideal laboratory, isolated and inescapable, for the study of human behavior.
> Escape is always possible on an island, desert, or mountain. We can always remove ourselves a little or a lot from the others, from what hurts us or bothers us. Not on a raft, [for] it is in moments of crisis and serious danger that we know one another best, when our true personalities come forth and present themselves without pretensions. This is when we

> express and act our real selves. This is the self we carry inside, that profoundly affects our lives but which on land we constantly hide because of shyness, education, convenience, sense of heirarchy, social fear, politeness, cultural-sexual inhibition, nationalism, or religious sentiment.
>
> —Santiago Genovés
> *The Acali Experiment: Five Men and Six Women on a Raft Across the Atlantic for 101 Days* (1980)

As William Willis had predicted during the *Age Unlimited* expedition, the drifting raft was eventually seen as "an ideal place to make experiments." (Willis 1966, 111) Willis, of course, did not mean "experiment" in the sense of a Heyerdahlian archaeological experiment, a test of recreated primitive technology. He meant that a raft could serve as an ideal platform for social experiment. Santiago Genovés, the Mexican social scientist from the *Ra I* and *Ra II* expeditions, followed up on this suggestion with something of an egomaniacal vengeance.

In 1973, as Vital Alsar led his fleet of three balsa rafts across the Pacific, Genovés with four other men and six women set out to test human compatibility on a raft voyage across the Atlantic. Given our current television environment, where ridiculous danger scenarios are played out with handpicked "survivors" in mild tropical island environments under the constant stare of cameras and paramedics, Genovés' *Acali* experiment now seems less like a bad 1970s situation comedy than a pilot for a millennial "reality" television series.

Genovés envisioned the *Acali* as a kind of "floating laboratory, a real sea adventure in which volunteers agreed to participate with the consequent risks, to obtain firsthand data about aggression, conflict, misunderstandings, and possibilities for harmony in a world of violence" (Genovés 1980, xv). The name of the raft came from a Nuhuatl, or Aztec, word, meaning 'house on the water.' The scientific nature of the experiment, however, was undermined from the start, when Genovés decided that he himself would be one of the humans trapped on the raft. This lent the experiment less than the full weight of isolation that it should have had. Besides setting up the conditions of the test, Genovés was himself, at forty-nine, more than ten years older than any of the other crew, and the only one with any experience on a drifting raft.

The six women Genovés selected for the voyage ranged in age from twenty-three to thirty-six. Only one was single. Of the others, two were divorced and one separated. The two American women on board were both married, one with two children and the other with three. The four men besides Genovés ranged in age from twenty-nine to thirty-seven. Two of the men were single; Genovés and two others were married. In his book Genovés gives all of the other participants pseudonyms. The potential for conflict seemed as geometric as the possibilities for romantic entanglements, and in fact Genovés' sociological study of the expedition reads like a self-congratulatory cross between a letter to a pornographic magazine and the daily racing form:

> Unquestionably, Santiago is considered the most intelligent on the raft and Ingrid is the one who inspires the least asexual love. Santiago should continue being the leader according to an overwhelming majority and the least suitable for being leader is Teresa. The persons most wanted alongside one in the cabin are Sofía, Santiago, and Marcos and as companions on watch, the same three... Most would like to make love to Santiago, Marcos, Esperanza, or Aisha. And the best friend, the one we think would be the best friend for the rest of our lives, is by an overwhelming majority Santiago... (ibid, 161)

Questionnaires filled out by participants during the expedition included such propositions as: "That everybody go around naked throughout one day" and "That a kind of fiesta be held in which everybody went to bed with everybody else" (ibid.). The former question received six affirmatives, the latter only four. There were no questions about one's feelings for the efficacy of the pre-Incan guara. Such scenarios as envisioned by the 'fiesta' caused the raft to be derided as the "sex raft." Genovés himself—especially after the voyage ultimately resulted in somewhat tamer sexual vistas than perhaps had been imagined for it—put forth a reason why his book about the expedition was translated into so many languages, more in fact than any rafting account since *Kon-Tiki*. "Freud wrote that the motor which moves the world is sex. Unamuno thought that it was envy. *Acali* has shown us that it is envy toward other people's sex." (ibid, xxii)

Because Genovés was seeking a floating laboratory, his raft was not any kind of recreation of prehistoric maritime technology. Designed by a British naval architect, Colin Mudie, the raft when finished looked somewhat like a multi-layered rectangular cake. Nearly forty feet long and twenty-one feet wide, *Acali* carried nineteen tons of water as ballast against the raft being capsized. Even in the latter event, two emergency exits were built into the hull. Genovés did not find nearly as much joy when he tried to locate psychiatrists and psychologists to help him analyze the experiment. "I cannot understand how people who have been at work professionally for only a few years are not interested in research unless they are paid for it, when, according to what they tell me, all of them earn four times more than I do as an anthropologist after twenty years in research" (ibid, 12). Despite the uncooperative colleagues, and the difficulty in obtaining permits and licenses for such a strange craft, Genovés had the raft in the water off the Canary Islands in early May 1973.

In the evening of May 12th, *Acali* was towed out to sea and let go. The first days at sea saw the raft advancing at a glacial twenty-seven miles a day, a huge worry for Genovés, because it meant that the raft would arrive in the Caribbean during the height of hurricane season. He was further surprised to learn that none of the volunteers bothered to read any of the materials he had passed out to them prior to the experiment. A few became interested during the voyage, some never did. The sexual relations that were so anticipated seemed to occur in fits and starts. Genovés speculated that, in the early days, it was because watching other people throw up from seasickness was not exactly an inducement to romance.

Later, partnerships formed despite, or perhaps because of, the odd number of men and women. Genovés, along with "Marcos," who was also one of his anthropology students, maintained a steady partnership with "Sofia," a separated, thirty-four-year-old French woman. Given this unusual triangle, there was little apparent interest in the sociological questionnaires in anything but couples. Genovés writes, "true communication among three people is difficult. It would be very French, but difficult!" (ibid, 113). One of the volunteers started to drink from a secreted bottle of alcohol, even though she was free to drink whenever she liked.

After a month at sea, *Acali* had drifted 1,430 miles from the Canaries and had another 1,800 miles to go before reaching Barbados. Beyond lay still another 1,500 miles to the Yucatan. The distance covered by the raft had increased to more than forty miles a day. The infighting amongst certain of the volunteers had increased as well, to the point when, a la *Survivor*, a secret ballot was taken to vote one of them off the raft as soon as it reached Barbados.

On July 19th, after two and a half months at sea, a towboat lurched out from Barbados to bring the raft into harbor in Bridgetown. Three days later, the raft was back at sea again, drifting across the Caribbean. The raft, as if sensing that the end was near, picked up speed, covering over seventy miles a day in one stretch. The Caribbean interlude was enlivened by an "experiment" in which each man spent an evening with each woman on a distinct and private section of the raft. Only three behavioral options were allowed during these times: silence, whispering, making love.

Before this particular experiment could proceed very far, Genovés was struck by appendicitis. Antibiotics kept the condition in check, and by the third week in August the raft was nearing Cozumel. As another tugboat towed another raft into port, Genovés himself seemed a bit bewildered by the meaning of his audacious experiment. What exactly had he explored? What kinds of data had he really produced? And would any of it seem relevant to his stated goal of reducing global tensions?

An answer of sorts was provided as Genovés was writing up his results. In April, 1978, Thor Heyerdahl and the crew of his third reed vessel, the *Tigris*, had managed to sail from the ancient cradle of civilization in Mesopotamia all the way to Africa, where local civil wars fueled by a global arms race halted their attempts to sail any further. In protest, Heyerdahl had his beautiful reed ship towed to sea and burned. Whether from the *Acali* experiment four years earlier, or the beginnings of civilization four thousand years earlier, it seemed that raft expeditions were proving that humans still had a very long way to go.

CHAPTER 12

THE OTHER EXPERIMENTS AFTER KON-TIKI: ADVENTURE

La Cantuta I
L'Egare I
L'Egare II
La Cantuta II
Pacifica
Celeusta

Abstract

One can trace the basic motivation of a good adventure to at least nine of the expeditions that followed *Kon-Tiki*. Of the expeditions whose motivation can be described as essentially adventurous, that is, with little or no serious scientific component, seven took place in the Pacific Ocean. Of these seven, five used some version of a balsa raft. These adventures were typified by William Willis' solo balsa raft voyage on Seven Sisters in 1954. Willis was certainly interested in the effects of prolonged isolation on the raft on the endurance of his body and mind. But during the 115 days it took him to drift from central Peru to Samoa, the 61-year-old American was mostly interested in recreating the feeling of being young, alive and healthy; in other words, of reliving the sensation of childhood adventure, only on a massive trans-Pacific scale.

Five other expeditions in the Pacific can trace their motivation almost exclusively to the desire for adventure. If William Willis was the most introverted and thoughtful of the adventurous raft expedition leaders, Eduardo Ingris was certainly the most colorful and cut the most carefree profile. In 1955, Ingris built a balsa raft named *La Cantuta* and drifted from northern Peru toward the Galapagos Islands. Ingris had assembled a crew that consisted of men from Argentina, Holland, and Peru, as well as a woman from Lake Titicaca, the high altitude lake and home of master reed boat builders. When the raft approached the Galapagos, it was snagged for three months in the Equatorial Counter Current. Unable to proceed east or west, Ingris and his crew had to be rescued.

Three years later, when Eric de Bisschop's raft *Tahiti-Nui II* landed at the Peruvian port of Callao in March of 1958, Ingris begged to be included in de Bisschop's crew for its planned drift to Polynesia. "Despite the unfortunate result of [his 1955 expedition], Ingris was so eager to begin again that he came back every day and tried by new and ingenious arguments to persuade us to take him with us." (Bisschop 1959: 133)

La Cantuta I (1955)

In 1955, a balsa raft named *La Cantuta* I drifted from northern Peru toward the Galapagos islands. The captain of *La Cantuta I* was "a cheerful curly-haired Czech" (Danielsson 1960, 133) named Eduardo Ingris. Ingris had assembled a crew that consisted of men from Argentina, Holland, and Peru, as well as a woman from Lake Titicaca, the high altitude lake and home of master reed boat builders. When the raft approached the Galapagos, it was snagged for three months in the Equatorial Counter Current. Unable to proceed either east or west, Ingris and his crew had to be rescued.

L'Égaré (1955)

Forshadowing the complete disaffection of the children of the second world war generation, Henri Beaudout, a young Frenchman living in Montreal, decided in 1953 that he had had enough of "adjusting [himself] to the vast rhythmic pattern of modern industrial and scientific life." (Beaudout 1957, 12) He yearned to do something, anything, to break away from the strictures of young married life in an increasingly meaningless world. To interrupt his ennui, he dreamed of founding a Canadian explorers' club, but had no qualifications to assume the lead of such an organization. One night the idea of a raft expedition entered his head, and would not leave when the sun came up. Thus one of the strangest post-*Kon-Tiki* raft expeditions came into being.

Beaudout conceived the idea of drifting from Canada to England on the Gulf Stream current on a "primitive" raft in order to enhance his bona fides for his imagined explorers' club. Consulting with furniture-makers in Montreal on the subject of the buoyancy of wood, all were in agreement that the wood Beaudout required for his Atlantic drift was red cedar, a tree harvested only on the other side of Canada, in British Columbia. The furniture-makers told Beaudout that red cedar "is composed of cells which are so little absorbent of water that after three months' immersion in sea-water...the logs of my raft would not contain more water than when first launched." (ibid, 15)

Unlike Bombard and Willis, Beaudout had already decided that he could not drift the Atlantic by himself, so as he tried to raise the money to build his red cedar raft he simultaneously cast about for a crew. A favorable newspaper article in a Montreal daily netted him exactly one volunteer, and even this individual backed out before even meeting with Beaudout.

Overcoming these intial setbacks, Beaudout managed to build his raft and round up two other crewmen in time to drift downriver from Montreal on June 11, 1955, just as Eduardo Ingris was fetching up in the Equatorial Counter Current off the Galapagos in the Pacific. On board *L'Égaré*, as Beaudout christened the raft, the three men drifted down the St. Lawrence seaway for two months, before the raft wrecked on the rocks on the coast of Newfoundland.

L'Égaré II (1956)

Returning home from the *L'Égaré* fiasco, Beaudout received a telephone call from one Marc Modena, a Montreal chef, asking to be considered as a new crewman if Beaudout ever tried the trip again. "Since babyhood [Modena] had dreamt of making a trip across some vast expanse of water, and I don't think fear could enter his mind" (ibid, 35).

Disheartened as he must have been, Modena's call cheered Beaudout somewhat, as did the fact that his red cedar logs had floated to perfection for sixty-six days. In the spring of 1956, Beaudout traveled with Modena, Gaston Vanackere, and a fourth crewman, to Nova Scotia. In Halifax, the men found eight massive trunks of seasoned red cedar. "These trees grow to a height of over 200 feet, and an old cedar, allowed to grow for decoration, would tower far beyond that; they grow only in British Columbia" (ibid, 36). Beaudout made a deal for the logs on the spot, and with his companions began to build a second red cedar raft, *L'Égaré II*, equipped with a square sail of Egyptian cotton and six wooden centerboards a la Peruvian balsa raft guaras.

The men invested all of their remaining funds in tins of food ("We wanted to cross the Atlantic, not stage a starvation test" (ibid, 48), and for 100 one-pint tins to carry fresh water. Beaudout thought the men could survive as well on an ocean full of fish once the raft crossed into the Gulf Stream, which he reckoned would take about six days. Besides forming a club, Beaudout now stated a second purpose for the voyage, namely to show that "a primitive craft, although only capable of sailing before the wind, could be navigated across the Atlantic from west to east with the aid of winds and currents." (ibid, 54)

On May 24th, 1956, the "primitive craft," with its tins of food, tobacco, and water, its radio and wristwatches, was towed twenty-five miles out into the Atlantic and let go. Almost immediately, *L'Égaré II* drifted into a twilight world of cold mists and contrary winds and currents. After a month of being pushed back to Canada as much as across the Atlantic, with no fish in sight, with two weeks of food washed overboard and rations down to a single potato and a single piece of toast per man per day, the fourth crewmember had had enough. The crew signaled for him to be taken off somehow, and a few days later a Canadian government marine research vessel came alongside and removed the crewman from the raft. The three remaining crewmen continued on their way, constipated and hungry but hopeful.

By the end of June, *L'Égaré II* was nearing the pull of the Gulf Stream, and the crew began catching both cod fish and black-backed gulls for food. The new diet apparently forced a series of vivid dreams and nightmares upon each of the crewmembers "pictures from our earlier lives in the most extraordinary detail" (ibid, 87). But the raft's progress was still only about twenty miles per day—less than half a normal individual's walking speed—and provisions were so low that Beaudout had resolved to attempt to buy food from the next passing ship.

When the next passing ship proved to be the U.S. Navy ship General Callan, Beaudout suddenly found himself in grateful possession of three weeks' worth of provisions courtesy of the primitive U.S. income tax code. Like true French connoisseurs, the crew critiqued the gift ("four chickens, not prime roasters—indeed, one had no wings" (ibid, 109)), but the birds as well as the rest of the gift were accepted nonetheless.

Soon after the navy ship departed, *L'Égaré II* moved into the warm water and warmer air of the Gulf Stream. Beaudout felt healthy enough now to climb to the top of the twenty-seven foot and inspect the seas ahead. He found the raft "not so much on the sea as in the sea, with a completely different horizon from that which even the smallest ship commands." (ibid, 111) Over the course of the next month, the raft was visited or passed by by several vessels en route to and from Europe. Beaudout noted that while the Dutch, the Scandinavians, and the Americans always stopped to give the raft its position and offer supplies, the British and the French almost never did so. While the raft's crew considered the latter ship's impolite, the austerity in both manners and gifts might instead be an indication of the relative post-war prosperity of the various countries.

By early August, seagulls returned to hover around the raft, a sure sign of land ahead. On August 11th, 1956, the raft came alongside a Breton fishing boat off the coast of England. The French fishermen provided the French rafters with wine, fresh bread, and fruit. The cuisine thus fortified, *L'Égaré II* drifted on, until on the afternoon of Sunday, August 19th, Beaudout picked out the flashes of a lighthouse fifteen miles away. When a Dutch passenger ship stopped to inquire of *L'Égaré II*'s status, Beaudout asked the captain to send a message requesting a small boat to come out into the English Channel and tow them in. A lifeboat was soon dispatched to take Beaudout ashore, where as a French subject arriving in Britain via a raft from Canada, there was a good measure of official paperwork to fill out. When Beaudout returned to *L'Égaré II*, the raft was being towed into Falmouth harbor, where a crowd of Britishers lined the quay to shout at them: "Well done, you crazy Frenchmen!" (ibid, 188)

L'Égaré II had drifted completely across the Atlantic from west to east, a distance of nearly 2,500 miles. That red cedar grew in British Columbia, on the other side of Canada, is of little importance in evaluating Beaudout's expedition as an Atlantic drift. No one suggests that North America indians logged red cedar in British Columbia only to carry it to Newfoundland and drift therefrom to Ireland or England (or Norway). Yet Beaudout had indisputably demonstrated that red cedar is exceptionally buoyant and an excellent raw material for long-distance transoceanic raft expeditions of at least two thousand miles. This is no small fact, given Thor Heyerdahl's insistence upon a drift migration route from the northwestern coast of North America to Polynesia via the Hawai'ian islands.

La Cantuta II (1959)

When Eric de Bisschop's raft *Tahiti-Nui II* landed at the Peruvian port of Callao in March of 1958, Eduardo Ingris begged to be included in de Bisschop's crew for its planned drift to Polynesia. "Despite the unfortunate result of this ill-planned attempt Ingris was so eager to begin again that he came back every day and tried by new and ingenious arguments to persuade us to take him with us" (ibid, 133). When at last de Bisschop decided against taking Ingris along, the Czech built another balsa raft. In 1959, *La Cantuta II* carried Ingris and a new crew from Callao all the way to Matahiva in central Polynesia.

Pacifica (1966-67)

Marc Modena, the jovial crewman from *L'Egare II*, joined another raft voyage in 1966, when he became chef and supply master on board the balsa raft Pacifica. Led by Spanish explorer Vital Alsar, the raft drifted away from the southern coast of Ecuador October 23, 1966. One crewman soon went crazy with fear, and had to be shifted to the first steamer able to return him to land. "Shortly thereafter we came ashore on the Island of San Cristobal, where we met some nuns who urged us to continue our journey, and several days later we set sail again, hoping to move westward along the northern extremes of the Galapagos" (Alsar 1973, 57). But like *La Cantuta* before it, the raft was caught north of the Galapagos islands in the deadly Equatorial Counter Current. After struggling with this shifting unpredictable current for 143 days, the crew sent out a series of distress calls as the raft finally began to lose buoyancy and sink. "We scrambled onto the roof of the cabin" Alsar later wrote, "as several sharks slithered across the sinking logs, praying for help as the raft dropped lower and lower" (ibid, 58). They were rescued on March 15, 1967, two hours before the *Pacifica* disappeared beneath the waves.

Celeusta (1969)

In 1969 the rubber raft *Celeusta*, captained by Mario Valli, drifted away from central Peru and landed on the island of Raroia, the same spot in the Pacific where the *Kon-Tiki* expedition landed twenty-two years earlier. In the same year as humans first landed on the moon, it was as if an exploratory spacecraft had been programmed by the same space agency to fly the same trajectory to the same destination.

Chapter 13

Tim Severin and Hsu Fu

> Naturally the day of our formal departure from Aberdeen was another 'gold day' in the Chinese calendar, selected by the feng shui expert as propitious for the undertaking. The expert … tinkled a little brass bell to attract the attention of the spirits, said prayers for our safety, burned paper offerings, and conducted a ceremony remarkably similar to the one we had already seen in Sam Son. Then we cast off and proceeded out of the packed harbour of Aberdeen, not to go directly to sea but to find an anchorage off the small outlying island of Po Ti where we could complete our last-minute loading of stores and wait for a favourable wind. I intended to learn from those medieval Arab travellers who, when they left on a long pilgrimage to Mecca, began with a very small step. Their first day's journey was never more than a mile or so from their town. There they stopped and made camp, so that everyone could discover what they had left behind … and run back home to get it. *Hsu Fu* was setting out on a voyage of perhaps 6,500 miles so it seemed a wise example to follow.
>
> —Tim Severin
> *The China Voyage: Across the Pacific by Bamboo Raft* (1994)

In the early 1990s, Irish explorer Tim Severin revived the old hypothesis that Chinese mariners might have reached the Americas two thousand years ago. Severin had been impressed when he met with the Cambridge University scholar Joseph Needham.

Needham's multi-volume work on Chinese civilization included strong hints of contact between China and the Americas.

One of the main artifacts cited by the proponents of transpacific contact is the centerboard. The same centerboard, or guara, that stabilizes the balsa rafts of Ecuador acts as a sliding leeboard for the rafts of Asia. There were two possibilities: the technology had been arrived at independently on both sides of the Pacific or, as Heyerdahl and many other speculated, prehistoric mariners carried the idea for the guara from Asia to Ecuador or vice versa. Severin concluded that the most likely method of contact between the two areas was the Chinese bamboo raft. In 1993, he set out to build one such extremely large bamboo raft and test his idea.

Severin traveled to Taiwan to see if the kinds of bamboo rafts mentioned by Needham still existed in the place of their birth. As he observed Taiwanese rafts, it is interesting to speculate whether Severin knew of Ling Shun-Sheng's article on "[The] Formosan Sea-going Raft and its Origin in Ancient China," published in 1956. Shun-Sheng's article describes in detail the construction of the centerboard-equipped sea-going Taiwanese bamboo raft, and makes a case for the mention of bamboo rafts in the Chinese historical record as far back as 1174 C.E., and legendary notes about rafts in Chinese literature as far back as 3,300 B.C.E. To Severin's surprise, he found the Taiwanese raft as common as they apparently were in prehistory, with two jarring exceptions. The bamboo of Shun-Sheng's rafts had been replaced by PVC plumber's pipe as the construction material of choice; and the outboard motor had replaced the sail as the method of propulsion.

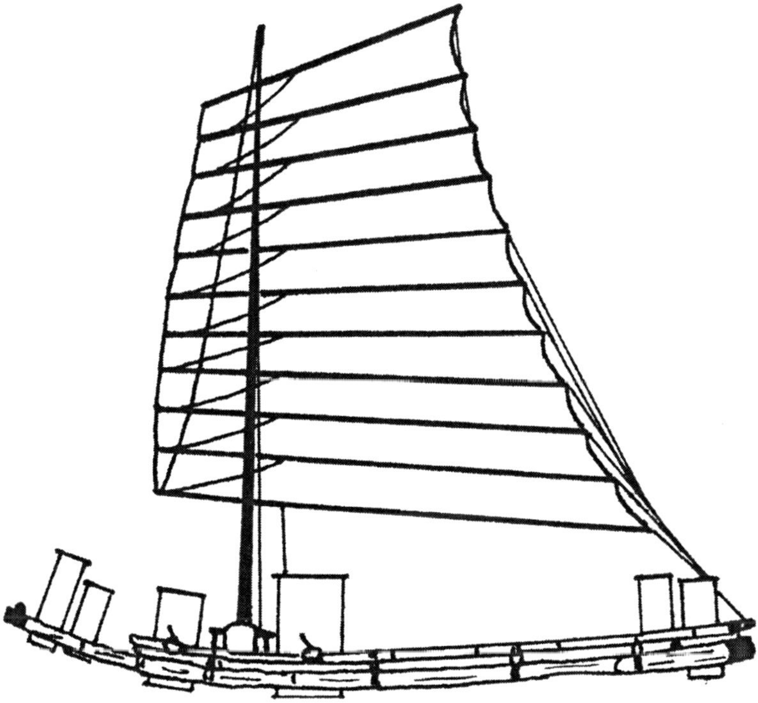

Fig. 18. Taiwanese bamboo raft (after Shun-Sheng 1956)

Forsaking the plastic rafts of Taiwan, Severin discovered that traditional bamboo rafts were still built in the communist nation of Vietnam, where a twenty-year economic blockade by the U.S. had left the fishermen utterly unable to afford luxuries like outboard motors or plastic plumber's pipe. In the coastal village of San Son, Severin collected enough information for the British naval architect Colin Mudie—the same man who had designed Santiago Genovés transatlantic *Acali* raft twenty years earlier—to sketch out a plausible bamboo raft for an attempt to cross the Pacific.

Of the hundreds of species of bamboo, Severin located the thirty-foot-long, six-inch wide bamboo stalks for his main hull in the interior of the country, along the Laosian border. Rightly terrified of predictions that any bamboo raft attempting a transoceanic crossing would be eaten by the marine invertebrate *teredo navalis*—the same organism that had literally eaten the experimental junk *Tai Ki* to the bottom of the Pacific in 1974—Severin took immense pains to try to protect his raft. The bamboo was harvested when the sap in them was the least, to protect against a possible infestation by insects that eat the sap and then disgorge eggs into the wood. His workers coated the hollow logs with layers of laquer, which produced the result of both protecting the raft and inflicting an allergic reaction to the lacquer on the raft-builders. All but two of the forty raft-builders were affected with swollen eyes and limbs, especially the ones who sought to gain supernatural protection against the lacquer by bravely licking it from their fingers.

The bamboo was lashed together with ropes made from rattan, then heated at selected points to give the raft a characteristic upsweep near the bow and stern. When finished, the sixty-foot-long raft was delivered to a harbor near Hanoi. The most curious aspect of the bamboo raft was that, like a balsa raft, waves did not travel over the hull but through it. This accounted for the raft's almost uncanny stability. But being much more flexible than rigid balsa logs, the effect on a bamboo raft was something like standing on board a twisting sieve in mid-ocean as waves swept through the decks, soaking everything in their path. It made for some interesting speculation as to how transoceanic mariners might have managed to keep themselves dry on voyages of 6,500 miles from Hong Kong to the Americas. Severin attempted to solve the problem by adopting the idea of the waterproof Vietnamese basket boat and adapting it in the form of two bamboo shelters on the decks of the raft.

In Hanoi, the raft was fitted with three chesnut masts supporting hand-stitched cotton and silk fan-shaped Chinese junk sails. Like the bamboo logs, the sails were treated with a natural substance—in this case one made of the boiled roots of an inedible yam—to ward off rot. A windmill charged a battery that powered a satellite radio. Severin would be able to transmit his position via fax each day. If the raft ran into trouble, rescuers would know with pinpoint accuracy where to search for the stranded crew. Severin named his raft *Hsu Fu*, after a lengendary Chinese mariner sent off on an expedition into the eastern oceans around the year 219 B.C.E. by China's first emperor Qin Shihuang. With the possible exception of Heyerdahl's *Ra II*, *Hsu Fu* was the most beautiful recreated prehistoric raft ever attempted.

Hsu Fu was transported to its departure point of Hong Kong on board the deck of a freighter. In Hong Kong, Severin learned that a virtual religion had developed over the years dedicated to the memory of *Hsu Fu*, whom many Chinese, Taiwanese, and Japanese consider not a mythical but an historical figure. The knowledge gave the recreated voyage even more meaning and impetus. On May 17th, 1993, with a crew of northern Europeans complimented by a single Vietnamese raftsman, *Hsu Fu* rather staggered out of Hong Kong and toward Taiwan, 120 miles away.

A sign of trouble appeared almost immediately. After all the care in treating the bamboo against *teredo navalis*, an almost unforgiveable lapse had occurred when bamboo used for the cabins had been selected carelessly. Once at sea, Severin discovered his cabins in the process of being eaten by beetles, and prayed the infection would not spread to the bamboo logs of the main hull. From the experience of *Tai Ki*, of Eric de Bisschop's *Fou Po*, and John Halsett's *Illa Tiki*, one could almost speculate that ancient mariners must have assembled any transoceanic raft in the prehistoric equivalent of a "clean room," like those used to piece together interplanetary spacecraft.

A week into the voyage, with the raft crawling along at little more than one knot an hour, the bamboo mainsail spar snapped and was replaced by one of the spare lengths of bamboo carried on board the raft. When the winds increased too much, the Chinese sails had to be folded down lest the masts snap. The constant twisting of the bamboo hull meant that one had to very careful not to let an arm or a leg slip between a temporary opening in the deck, lest one risk amputation. Other than these weaknesses, Severin was astounded at how the raft responded in a gale. Where another ship would be rolling and pitching, *Hsu Fu* remained completely flat and stable in high winds and waves, allowing big rollers to wash right through the decks.

As May turned to June, and the progress of the raft was slow, Severin maneuvered *Hsu Fu* around the southern coast of Taiwan in an attempt to pick up the Japan Current. This twenty-mile wide river of fast-moving water was the main "cultural current" of so many Asia to America contact theories. And Severin found himself in the middle of one of theorist's favorite scenarios. The Japan Current was carrying the raft away from Taiwan and, with rapidly diminishing supplies on board, pushing the vessel toward Japan itself. A helpless crew, drifting on an out-of-control raft. It was one of the very possibilities Heyerdahl and others had advanced for half a century.

After avoiding pirates, meeting up with a friendly yacht (which took off one crewmember and replaced him with an another,) and snapping the foremast in a high wind, *Hsu Fu* managed to sail into a harbor in the Japanese Ryukyus islands on June 12th. In a layover of nine days, two damaged masts were replaced and a new crewmember taken on board. As the raft made its way along the Japanese coast amid the worst and wettest weather in forty years, Severin noted that the bamboo deck seemed to be losing its buoyancy. Moreover, the continual twisting and turning of the entire deck structure served as a constant strain on the rattan lashings.

In the middle of July, the raft called at the Japanese port of Shingu, where the *Hsu Fu* of history was said to have landed, and where a shrine had been built in his honor. Upon leaving, the raft ran into a typhoon, and Severin was bedeviled in his attempts to convince any of the fishermen of a nearby port to help tow the raft or even offer it shelter from the storm. At Shimoda, the *Hsu Fu* was stocked with provisions for the transpacific attempt. The raft had come 2,000 miles from Hong Kong, and 4,500 miles now separated it from America.

With a crew of five, *Hsu Fu* drifted away from Japan and out into the North Pacific on August 5, 1993. Loaded with provisions, the raft rode so low in the water that waves washed right through the cabins. Valves that Severin had built in Vietnam to allow seawater to drain from the watertight cabins had rotted, allowing the sea to slosh directly into the sleeping areas with each rise and fall of the raft. Nothing on the decks remained dry. It seemed like an inauspicious way to begin a voyage across an entire ocean. Most raft voyages did not have their decks awash until vast stretches of open ocean had been crossed. By choosing to sail from Hong Kong, Severin had already subjected his raft to as much soaking and punishment as *Ra I* had endured when it sank out from under Heyerdahl in 1969. And Severin still had nearly 5,000 miles in front of him.

The raft was tested almost immediately, when *Hsu Fu* skirted along the outer path of a typhoon toward the end of August. Two weeks later, a third of the way across the Pacific, Severin himself was knocked down by a swaying mainsail boom and broke two ribs. As his ribs mended, Severin calculated the raft's movement eastwards. Originally, he had thought the raft could make the crossing to America in roughly ninety days, with an average daily run of fifty miles. But *Hsu Fu* was crawling along at forty miles a day, a pace that would add another month to the voyage, and add a month to the strain on the bamboo. The rattan rope was rapidly being used up, and Severin doubted whether enough remained to finish the crossing. In late September, halfway across the Pacific, he confided his doubts to his private journal:

> Now I am aware that this present voyage is far longer, across more open water, in high and stormy latitudes, aboard a vessel half-sunk before it begins, and subjected day and night to flexing, twisting, and battering of thin-walled bamboo tubes—a grass not even a timber—ropes made of flimsy fibres, masts held up by jungle vines. (Severin 1995, 222)

As the raft drifted to within 2,000 miles of the west coast of North America, it became clear to Severin that there was little real danger of the bamboo raft sinking. The true threat lay in the raft's constant motion, which was slowly working the intricate rope lashings into dust. The rattan that tied the bamboo raft together was not up to the constant strain of nearly six months at sea. Severin told his crew that if he decided that the raft was no longer safe, he would call for rescue and abandon the experiment. At one point, the U.S. Coast Guard cutter Jarvis, en route from Alaska to Hawai'i, layed-to near the raft, while a boarding party checked the condition if the *Hsu Fu*. Severin was heartened to note that the survival gear he had insisted on putting on board *Hsu Fu* "was probably of a higher standard than that carried by the Coastguard" (ibid, 262).

Yet the raft itself was beginning to deteriorate before their eyes. Large main bamboos began to work loose from the hull and trail lazily astern. One thousand miles from the American coast, and nearly 4,000 miles from Japan, the crew watched as several bamboo logs dislodged and drifted away. The raft was breaking up underneath them. After more than 100 days at sea, the crew now noticed that teredos were finally making their appearance as well, eating their way through the bamboo.

With the crew's spirits flagging and his raft breaking up, Severin in early November gave the order to bring the experiment to a halt. As he wrote in his journal: "If Chinese mariners made this voyage in ancient times, they came ashore in pre-Columbian America so exhausted they would have been on their knees" (ibid, 277). On November 12th, a passing container ship picked up the crew, but not before they set all sails, and prepared the raft to drift on unmanned. Severin tied a note to the raft asking anyone who found it to report its position to the Mariners' Museum in Virginia. But *Hsu Fu* was never seen again.

More than even the voyage itself, Severin demonstrated through the voyage of the *Hsu Fu* an acute awareness of dozens of seemingly minute yet intensely interesting details attending transoceanic raft expeditions. Details involved in construction, in sailing, in endurance, in crew reactions and the receptions afforded, or not afforded, to the vessel en route, all of these points of analysis were kept constantly in mind throughout the Severin's narrative. What prehistoric treatments might have afforded various woods protection from teredos? How would a weakened crew have brought a crippled vessel ashore with no help from completely exotic strangers on shore? Almost nothing is known about these aspects of prehistoric voyaging in any kind of experimental setting.

Severin is further to be lauded for including an artist at every stage of the project. This inclusion gained a unique perspective on the experiment, one that is usually remote from anthropological grasp. With his voyage of 6,500 miles on a bamboo raft, Tim Severin had set a new standard for all future archaeological experiments.

CHAPTER 14

THE ORIGINS OF THE RAFT

The Origins of the Raft
The Origins of the First Rafters

…the original Negritic immigrants to Greater Australia … reached the Sahul Shelf south of present-day New Guinea and so had equal access both to the high forested landmasses to the north and to the drier, more lightly wooded portions of the Shelf leading directly into present-day continental Australia… [This begs a question] concerning the dynamics between an established continental population intermittently assailed by raft loads of incoming peoples of different populational character. It is easy to suppose that a second wave of immigrants across the islands to the Sahul Shelf would have been quickly and simply absorbed by those already in possession of Greater Australia. But the present distribution of peoples in New Guinea, as contrasted to Australia, indicates that it could not have happened that way.

—Joseph B. Birdsell
"The Recalibration of a Paradigm for the First Peopling of Greater Australia" (1977)

The Origins of the Raft

Although not a drift expedition, the voyage of the *Tai Ki* in 1974 revived long-standing speculation that the ancient civilizations of China had influenced—if not outright germinated—the great cultures of Pre-Columbian Central America. An Austrian journalist named Kuno Knöbl and a team of explorers initiated a project to build a sailing vessel on the presumed model of a Chinese junk around 100 C.E. The dating of such an expedition, of course, would have placed Chinese mariners in Central America just at the start of the greatest period of cultural development of the ancient Maya.

Knöbl and his associates built a twenty-meter long wooden junk and sailed in it from Hong Kong on June 18, 1974. By the middle of August, various health problems forced Knöbl himself to abandon the ship. The rest of the crew continued on, as *teredo navalis* worms attacked the vessel with the same viciousness with which Eric de Bisschop's *Fou Po* had been destroyed forty years earlier. After sailing and drifting for another two months to a point about 700 miles south of the Aleutian Islands—and still 2,000 miles from North America, the *Tai Ki* was set adrift in October and the crew rescued by a passing merchant ship.

The idea that Chinese mariners could have crossed the Pacific two thousand years ago leads one to consider the very origins of maritime travel. Could late Pleistocene-era bamboo rafts have drifted out from the low-lying settlements of continental Asia and across to the northern coast of New Guinea. Who was the first human to take to the sea? What did their vessel look like? Where were they going? Did they know—could they see—where they were going? How would one explore this idea experimentally?

Thoughts of such remote prehistoric raft expeditions came forcefully to me during a week in Indonesia in 1997. Surveying the harbor at Manokwari, a small port situated in the Vogelkop (Bird's Head) sector of New Guinea, one could sense the almost crushing weight of the many layers of human history and exploration that have settled over the Southwest Pacific. The canoe we used for our bathymetric survey in search of a sunken American flying boat glided effortlessly over a harbor visited in the nineteenth century by Alfred Russell Wallace during his research into the evolution of beetles. The double-outriggers fashioned from thick bamboo logs barely kissed the water before their natural buoyancy lifted them back above the surface.

On our flight to the port city, one could look down upon an ocean of bamboo rafts and double-outriggers, like water bugs alight on a pane of glass. It was simple to imagine that the transport plane was really a time machine, and you were suddenly flying over a prehistoric human colony from twenty-five millennia in the past. During evenings alone in my room, I pored over histories of this swampy backwater of human history and prehistory.

New Guinea, occupied by humans at least 10-20,000 years before the earliest human groups arrived in North America, went unseen by Westerners until 1606. When Magellan crossed and named the Pacific in 1519-20, his expedition missed seeing the Vogelkop by a mere hundred miles or so, when it passed west of New Guinea en route home from the Philippines. Magellan's expedition, like dozens after it, was seeking in part an undefined Terra Australis Incognita, an Unknown Southern Land. This landmass had been hypothesized since the time of the Pythagoreans, six centuries before Christ, as a counterbalance to the known landmasses of the northern hemisphere. Maps purporting to show the undiscovered country were printed as early as 1529. The continent gripped the imagination of geographers and explorers alike for two hundred and fifty years, until the meticulous expeditions of Captain James Cook in the 1770s showed that the Southwest Pacific consisted of Australia, Antarctica, and hundreds of other islands of various sizes.

Between the eras of Magellan and Cook, Europeans largely ignored New Guinea and the Southwest Pacific. The Portuguese and Dutch fought over the famous Spice Islands of Banda and

Ternate (which, with Western New Guinea, would later become parts of Indonesia). The Spanish gained control over the Philippines in 1565, and established a transpacific route home from Manila via Acapulco, overland to the Caribbean, and then across the Atlantic. The British, supporting raiders such as Drake, Cavendish, and Hawkins, pirated Spanish settlements and ships from Peru to the Indies and back again.

Although by far the largest island south of the equator, New Guinea existed in almost complete isolation from the outside world until well into the nineteenth century. With adjacent footholds in Batavia (Jakarta), the Dutch and British sea-borne empires competed for the wealth of the Indonesian islands save for New Guinea. With most of the riches laying to the west, neither colonial power made much of an attempt to explore or settle the vast island. Even today, in an age of maps created by satellites, detailed maps of the interior of New Guinea frequently carry the hopeless notation "unexplored." Previously unknown tribes were being discovered in the interior of New Guinea as late as 1969.

The Dutch asserted vague rights over western New Guinea as early as 1714, but did not formally claim this western sector for more than a century, in 1828. If in fact a Dutch vessel from the 1830s rests at the bottom of Manokwari Harbor—as a local communicated to me during our 1997 survey—it represents a pivotal shipwreck in world history, existing at a boundary of Dutch control and British insurgency into the area. The British built a fort at Dore Bay near Manokwari as early as 1790, but starvation and beri-beri killed so many that the place was abandoned within half a decade.

By June of 1885, after years of colonial anguish and wrangling over "interests," the island had been divided between the Dutch, British, and Germans. Ten years later, with typical indifference to the natives, the Dutch had consolidated their claim to the parts of New Guinea west of the Fly River, while the British and Germans had divided the eastern half of the island between them. If the areas of the coast were little known to the Europeans who claimed them, the interior of New Guinea might as well have been on the backside of the moon. The natives of the interior mountain ranges were known as scattered, various, without common culture or appearance, and inviolately hostile to outsiders.

Such boundaries as Europeans might draw held little or no meaning to the native peoples. No European had more than a vague idea what lay in the interior. Such knowledge as existed consisted of occasional reports of Dutch naval vessels cruising along the coastal settlements; the odd scientist on an episodic investigation; the ubiquitous traders; and a few early missionaries. The Protestants of the Utrecht Society established the first missionary settlement at Manokwari in 1855. There, the wiry men of the Arfak Mountains—the tribes of the Moiray and Meach; and the Hattam we met at Minyambau during a flight into the interior —bore reputations as the fiercest headhunters in western New Guinea. They wore their ancestor's skulls as charms. By one account, with rifles dropped by Allied aircraft during World War II, these tribesmen both fought and then ate the Japanese.

Three years after the arrival of the Protestant mission, Alfred Russell Wallace explored nearby Dore Bay for five months collecting the birds and insects central to his understanding of biological evolution through natural selection. The interest of his fellow Englishmen in New Guinea obtained primarily on the southern coast, and that from a desire to ensure naval and commercial access to the Pacific through the Torres Strait. A Royal Navy captain by the name of John Moresby surveyed the southern coast in the 1870s, giving his family name to the best harbor on the southern shore. Germans approached the island from the north, trading in copra, and trochus and tortoise shells.

Colonization plans abounded, centering on notions of gold and spices thought to be readily gathered in the mysterious interior. Pressure to colonize New Guinea came from lawyers in London and speculators in Sydney, based on the seemingly timeless ability of westerners to believe in the presence of gold mines across the tropics. Expeditions to the island returned with spectacular scientific collections, but with nothing that the exploiters thought valuable enough to stimulate an economic rush. Especially discouraging were the low-lying coastal swamps, the ever-presence of "fever" (malaria as such had not been identified as yet), and the seeming impossibilities of penetrating into the presumably healthier interior.

The first true occupiers of New Guinea since the Pleistocene ice age turned out not to be economic exploiters but religious fanatics. Traders had led the Zealots of the London Missionary Society to believe that New Guinea offered a fertile field for a modern crusade. These British missionaries were in equal parts spearhead and witnesses to the British attempt to annex southern New Guinea.

Archaeologists have made several inroads into the culture history of the natives of New Guinea. By various reckonings, primitive humans arrived in New Guinea at a time when it was connected by a land bridge to Australia some 40,000 years ago, which accounts for racial similarities of some New Guinea tribes with those of the aborigines of northern Australia. Whenever sea levels dropped during periods of worldwide glaciation, New Guinea and Australia became connected, and formed a continent prehistorians refer to as Sahul. For the Stone Age tribes of the interior—as we could see for ourselves as we crossed the highlands of the Vogelkop—little has changed since Sahulian times. In fact, modern anthropologists have come to believe that these peoples resemble what all equatorial peoples looked like and how they lived 50,000 years ago.

Trade between the islands of New Guinea and Asia has been going on for at least four thousand years, perhaps as many as 35,000 or more. Cloves and bird of paradise skins have been found in the Middle East and dated from the former date. Initially, this trade grew through the traditional inter-island trade networks. Five thousand years ago the north coast of New Guinea was the periphery of a pottery-making complex resembling that made in Asia. This presumably indicates the existence of trade with Asia. By 3,600 years ago, a new pottery tradition erupted in New Guinea, known as Lapita. This style of pottery spread as its makers colonized the western and central Pacific. However, this tradition ceased with the arrival of specialist traders 2,000 years ago.

These specialist traders worked within traditional kinship-based trade networks that exist to this day. Evidence for these trading networks are found in the arrival of Indonesian bronze and glass artifacts in New Guinea. These trade items were exchanged for Bird of Paradise skins and feathers, which were much prized in Asia. The desire to possess these feathers and skins drove the trading cycle with the coastal settlements of New Guinea.

Around the year 300 C.E., one can trace a decline in interest in plumes and a rise in the spice and forest product trade. The local knowledge gained during plume trade let traders exploit already established contacts. Although most of this trade occurred in what is now Indonesia, it drew traders from throughout Asia in search of these riches. Chinese traders reached the Spice Islands of Ternate and Banda in the 13th century and gradually expanded their influence over the next 200 years. Traders extended their search to northern New Guinea by the 15th century.

It was this trade that drew the Europeans: the Portuguese in 1511 captured Melaka in an attempt to monopolize the trade; later the Dutch intervened in local conflicts, disrupting local economics and politics. Increased production brought a spice glut by 1630; the price of cloves declined and the Dutch introduced trade restrictions. In 1653 the Dutch began paying subsidies to the Sultan of Bacan and in 1657 to the Sultan of Tidore in lieu of foregoing their spice profits.

As trade increased, locals began to feel that they were not sharing equitably in the profits from trading and fighting ensued. Bronze spearheads and daggers from the Jayapura region attest to confrontations with traders beginning 2,000 years ago. In historical times, Portuguese traders established a post in Ternate in 1522, only to be driven from the area by 1574. The Bandanese traders were able to exclude Europeans until 1621, when the Dutch defeated them and established trade relations in this region.

As the prices for spices and forest products fell in the late 1600s, another disruption occurred in Asia, the Manchu invasion of the Ming controlled provinces of China in 1640. Although trade was disrupted for forty years during this war, trade with New Guinea eventually reopened in 1684. This trade now concentrated on marine and jungle goods like bird nests, mother-of-pearl, and edible sea slugs known as beche-de-mer. The trade prompted alliances between the Chinese and the Dutch and led to the expansion of trade posts throughout this region. Makasar and Bugis traders scoured the area reaching most of New Guinea and northern Australia.

With the intention of capturing this trade, the British founded Forts Dundas, Wellington, and Essington in northern Australia from 1824 to 1849. The Dutch countered by claiming all of the southern coast of New Guinea as their territory by claiming it for the Sultan of Tidore. All of these settlements proved failures, and this trade declined in the 1890s.

In 1848, the Dutch used the Sultan of Tidore to extend their claim to the north of New Guinea. In reality, the Sultan had little influence and the area was controled by the Dutch. One of the new Dutch stations was established at Manokwari, a port that had long been the trading center of northern New Guinea. Soon after, whalers and others in search of resources that had been wiped out elsewhere began to arrive. Investors, settlers, and missionaries followed, but by 1900 industry had declined to a few copra plantations and local trading. Plume hunting ceased in 1926, due to a combination of a decline in the plume fashion craze and government prohibitions aimed at conserving remaining bird stocks. New Guinea returned to the backwaters of Pacific trading until the gold fevers and modern mineral extraction expeditions of the 1920s. The same rugged mountains and swamps that had isolated the interior since the last ice age, had also hidden the real richness of the islands until modern methods of travel and extraction made mining the interior profitable.

While confusing, the European arrivals and influences in New Guinea can be traced fairly exactly. The timing and effects of prehistoric landings in New Guinea are other matters altogether. It is thought that there was no great mass migration of early humans and proto-humans into Indonesia. Rather, primitive peoples arrived in fits and starts throughout this enormous archipelago over the course of forty millennia. Through recent archaeological excavations, *Homo erectus*, long thought to be extinct nearly half a million years ago, may have been alive as recently as 25,000 years ago on Java. If so, they likely competed for space and resources with modern humans. It seems clear that Java was one of the earliest areas in which humans and proto-humans sought to live, at a time when the island was part of the Asian landmass and most of Europe was covered by continental sheets of ice.

The fossil skull of Java Man (*Homo erectus*) was first found in Trinil in Central Java in 1891. Charred remains of bones seem to attest to the controled use of fire. The species either could not adapt to the environmental conditions or was forced into extinction by the newly arrived Homo sapiens, ancestors of the present-day Indonesians. Beginning about 40,000 years ago, the sapiens began a movement from Australia into New Guinea and the Lesser Sunda Islands. Others arrived in western Indonesia from southern China some 30,000 years ago. Deep in the jungles of Irian Jaya one can find the genetic descendents of the Negritos, pygmy-like humans who also spread through Indonesia beginning 30,000 years ago. Skulls found at Wajak in East Java are evidence of the first ancestors of the present-day Indonesians, the Homo sapiens who populated the area 10-12,000 years ago.

How these early peoples voyaged over fairly long stretches of water is a matter of intense interest to archaeologists. J.B. Birdsell identified five different routes from the ancient Southeast Asian continent of Sunda to the combined Australian-New Guinea continent of Sahul (Birdsell 1977). It seemed probable to Birdsell that "the watercraft used in the late Pleistocene were superior to those found in recent times in Australia and Tasmania. It is highly probable that there was a constant if somewhat straggling trickle of small groups of human beings over all or most of the routes. The size of the watercraft likely to have been used suggests that the groups consisted of small biological families" (ibid, 123).

Birdsell identified thirty-five different species of bamboo along these routes, indicating both the possible raw materials that prehistoric mariners may have used to construct rafts and the paths from which they were set adrift. Such varieties of both routes and raw materials provided scores of testable hypotheses. As Birdsell noted, assuming these many species of bamboo existed during ice age times of lower sea levels, they offer "a useful role for experimental archaeology" (ibid, 144). One could readily imagine a fleet of different bamboo rafts exploring dozens of possible routes for early humans from Asia to Sahul. Chapters 39-41 describe the two documented cases—one from real life and the other from the world of archaeological experiment—when bamboo rafts attempted to drift "out of Asia."

The Origins of the First Rafters

During an archaeological expedition to the Indian Ocean archipelago of the Maldives Islands in the mid-1980s, Thor Heyerdahl spotted a large, abandoned bamboo raft in a lagoon at the island of Viringili. This particular type of raft was foreign to the Maldives, and a group of children were playing on it. The locals told Heyerdahl that some fishermen had discovered the raft the day before, on the outer reef off the east coast of the island. Because it carried no crew, the fishermen were suspicious and tried to burn it.

Obviously, it was no use to try and keep Thor Heyerdahl away from exploring a raft that had just drifted from some distant shore. He arrived at the bamboo vessel just as the boys were tearing it apart to sell the pieces at a local auction. The children jumped overboard as the adults arrived to inspect the apparition.

"It was a raft indeed: large, and built from giant bamboo. I had never seen bamboo as thick as this. Thick as telegraph poles and tied together in three layers under eight cross beams of wood with another layer of thick bamboo as a deck well above the water... Twelve meters long and three wide (39 feet by 10 feet approximately) with a freeboard of 40 centimeters (16 inches), the whole structure recalled the balsa raft *Kon-Tiki* which had been large enough to carry me and five companions across the Pacific" (Heyerdahl 1986, 188).

Seventy-year-old Heyerdahl then boarded the raft, which did not rock or move under his feet, and almost immediately he felt transported forty years into the past. Though badly burnt, the bamboo cabin reminded him of the small space he had shared all those years earlier with Danielsson and Haugland, Raaby and Watzinger and Hesselberg. The cabin floor was raised up some eight inches above the main bamboo deck, and a baked clay cooking area occupied one corner. The cabin roof was covered with a watertight layer of coconut thatch.

Making his way down the length of the raft, Heyerdahl encountered a few dried fish, which he thought had probably jumped on board themselves as the raft drifted on the open ocean. The only clues as to the origins of the raft were two paper labels, which he later identified as Burmese candy wrappers. The raft had most likely therefore floated nearly 2,000 miles from Burma, through the Bay of Bengal, and into the Indian Ocean to the Maldives. From his own rafting experience Heyerdahl looked at the barnacles growing on the bamboo and estimated that the vessel had been at sea about two months, pushed ahead of the Northeast Monsoon.

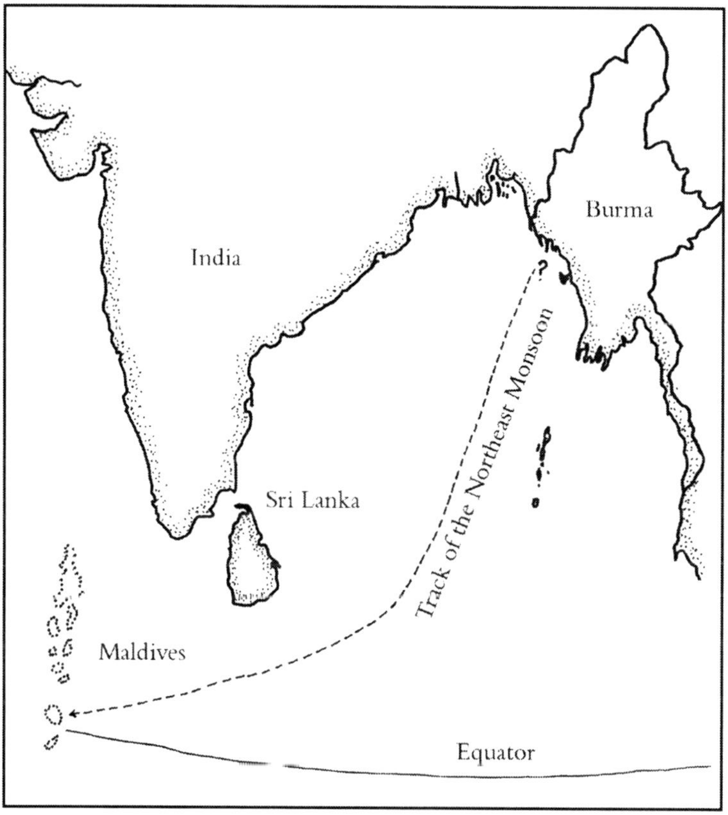

Fig. 19. Possible route from Burma to the Maldives of the bamboo raft Heyerdahl boarded at Viringili in the Maldives.

After a lifetime of drifting on primitive rafts, the sight of this Buddhist raft in the Maldives triggered much interesting speculation. Heyerdahl concluded that the teachings of Buddha had likely made a similar drift to the Maldives—a thousand years before Moslems settled in the islands—travelling from Nepal down the Ganges to the Bay of Bengal, to Burma, and across the Indian Ocean.

But what of the ultimate question? Had the cradle of civilization been constructed by farmers or fishers? If the latter, then was it made from reeds, or balsa, or bamboo?

CHAPTER 15

NOTES AND FURTHER READINGS AND STUDY

The increasing speed of and human dependence upon technology, with its seemingly concomitant loss of the natural environment and human connections to it, are themes reflected throughout the life and writings of Thor Heyerdahl, in such works as *Fatu Hiva: Back to Nature* (Garden City, New York: Doubleday, 1974), and *Green Was the Earth on the Seventh Day* (New York: Random House, 1996). As he writes in *Green Was the Earth on the Seventh Day* (p. 306):

> We have invented computers, fax machines, and faster ways to save time. But the time we have saved we need in order to earn enough to pay the bills for all these timesaving devices we did not need before. And we have become more stressed and more pressed for time than any generation before us.

Introduction

The James Cowan quote appears on p. 5 of *A Mapmaker's Dream: The Meditations of Fra Mauro, Cartographer to the Court of Venice* (Boston: Shambhala, 1996).

The *Kon-Tiki* Expedition

Recent enthnobotanical research shows that, of seventy-two species tentatively classified as intentional introductions into

Polynesia, only three, and maybe only two, are seen as arriving in Polynesia from South America (Whistler 1991). Moreover, the evidence for chili peppers on Easter Island is discounted because of an error in translation of an original Spanish text (ibid). Whistler admits that the sweet potato is indigenous to South America, yet remarks that it "was aboriginally introduced into the Pacific islands, possibly by Marquesans, centuries before European contact" (ibid, 51).

This, of course, begs the question of how the Marquesans obtained it to be in a position to spread it to the rest of Polynesia, or how the Peruvian name kumara followed the sweet potato across Polynesia. Heyerdahl's solution is well known, of course. Finney suggests that it is possible for a double-hulled canoe to make the voyage to South America, but the trip would not necessarily have been a pleasant one. It is likely that such navigation could only be accomplished with sustained blows of episodic westerly winds, perhaps during an El Niño event or propitious winter weather. "Working that far eastward with spells of winter westerlies would require a fortuitously long series of such events, and going far to the south to pick up the midlatitude westerlies would have exposed the canoe and crew to the chill, rough seas, and boisterous winds often found there…" (Finney et. al. 1994, 285).

Heyerdahl has commented extensively on his reasons for choosing to build a raft of balsa wood for his the *Kon-Tiki* experiment (Heyerdahl, 1950, 30-35; 1952, esp. 585-620; 1971, 17; 1979, 193-195, 201-218). He has simultaneously compiled an impressive case that pre-Incan Mochica mariners on both Lake Titicaca and coastal Peru navigated large and sophisticated ships constructed from totora reeds, the American fresh-water reeds so similar to those discovered in the three Easter Island crater lakes of Rano Raraku, Rano Aroi, and Rano Kao. But at the time of his decision to built a log balsa rather than a reed balsa, Heyerdahl, who had rejected prevailing doubts about log rafts, accepted prevailing opinion that reed boats could never withstand transoceanic crossings. He thus built *Kon-Tiki* from balsa logs and not totora reeds. (Heyerdahl, 1971, 17)

For a discussion of the early history of experiments in archaeology, see: Bruce G. Trigger, *A History of Archaeological Thought* (Cambridge: Cambridge University Press, 1989, 7, 61, 82, 86, 98, 271, and 399).

The scholarly voices raised against Thor Heyerdahl were not completely vituperous. At the 1960 Pacific Science Congress in Honolulu, the stage seemed set for a dust-up between the sailors and the drifters. The dean of Pacific archaeology, Kenneth Emory of Honolulu's Bishop Museum, appeared with Heyerdahl on a radio show, apparently on the expectation—or the hope—that they would come to blows over the seemingly eternal "question of Polynesian origins." But the confrontation fizzled. Both possessed too much class to devolve to shouting at one another—instead they agreed to disagree. Heyerdahl stuck with his position that the first people to reach the islands did so on rafts. Emory remained convinced from his own pioneering archaeological work that these populations arrived from the west (see Bob Kraus, *Keneti. South Seas Adventures of Kenneth Emory*, Honolulu: University of Hawaii Press, 1988, 376-377).

At a conference in San Francisco in 1963, Emory allowed that raft voyagers from South America could not be discounted as the agent who brought the sweet potato to Polynesia, nor could they be discounted as having exerted an influence on the prehistoric development of Easter Island. These were indeed the generous comments of a world-class scholar, one whose whole career had centered on excavating evidence that led to very different conclusions.(ibid, 389)

Something of a consensus on the subject has emerged over the past twenty-five years, with the discovery of a pottery-making culture known as Lapita. The Lapita people, so-called because of the first site where evidence of them was unearthed, have been radiocarbon dated to the period from 1,600-500 B.C.E. This would place them in eastern Melanesia, at the entrance to western Polynesia, at least a millennium prior to the possible drift into eastern Polynesia of pre-Incan Moche mariners from coastal Peru. (For a description of the Lapita complex see: Patrick Vinton Kirch, *The Lapita Peoples: Ancestors of the Oceanic World*, Blackwell, 1996.)

These dates, however, still leave empty the vast field of possible human sea-borne migration patterns during the last Ice Age, otherwise known as the late Pleistocene era, a period of time from roughly 35,000 to 10,000 years ago. Reduced sea levels—at times lower by more than one hundred meters—would have produced increased landmasses, profound changes in winds and currents, and a geometric increase in potential archaeological sites. It will take an enormous submarine archaeological reconnaissance to sort out what was going on during the late Pleistocene. As the brilliant Rhys Jones writes: "Some water straits were never closed during the [late Pleistocene] and to get to these islands, men had to cross various distances of sea. These acts of colonisation are of intense interest and we have to consider such variables as the rate of successful crossings; the probabilities against the survival of individuals in new ecological conditions; and more fundamentally, the probabilities against the survival of viable populations when the number of colonists at any one time is extremely low..." (See Rhys Jones "Sunda and Sahul: An Introduction," in J. Allen, et al., Sunda and Sahul: *Prehistoric Studies in Southeast Asia, Melanesia and Australia*, London: Academic Press, 1977, 7).

For critiques of archaeological experiments, see: Claassen, Cheryl "Answering Our Questions with Experiments," in Richard A. Gould, ed. *Shipwreck Anthropology*. (Albuquerque: University of New Mexico Press, 1983). For an overkilled and highly selective discussion of the imagined faults of the *Kon-Tiki* expedition itself, see: Bahn, Paul, and John Flenley, *Easter Island, Earth Island*. (New York: Thames and Hudson, 1992, 38-68).

De Bisschop's Response to *Kon-Tiki*

Apropos of the presence of bamboo as an industrial cultigen in the Marquesas, it is interesting to note Ferdon's work. Using early ethnohistorical accounts, he notes that bamboo in the islands was used for houses or huts; for knives for butchering (when split to have an edge); as burial scaffolding; for outrigger or outrider floats; even for trumpets (with a small tube emplaced at an angle in the larger tube. Reflecting de Bisschop's feeling that most maritime technologies had been long forgotten by the time of European contact, nowhere does he mention bamboo in use for rafts.

Ben Finney made a similar point at the 1999 Pacific Maritime History and Archaeology Conference in Honolulu. When I asked how it was that the Polynesians lost their deepwater skills, Finney pointed out that there may be a hundred different reasons on a hundred different islands. For example, large trees for canoes were wiped out on Easter Island, so even if the idea for a canoe managed to persist in the culture, the raw materials to make one had not.

De Bisschop would very likely have known of seagoing bamboo rafts from his years in China. It is therefore interesting to speculate whether, while building *Tahiti-Nui*, he read Ling Shun-Sheng's excellent article on "[The] Formosan Sea-going Raft and its Origin in Ancient China," Bulletin of the Institute of Ethnology Academia Sinica, (1, 2-54), published in March of 1956. Shun-Sheng's article describes in detail the construction of the centerboard-equipped sea-going Taiwanese bamboo raft, and makes a case for the mention of bamboo rafts in the Chinese historical record as far back as 1174 C.E., and legendary notes about rafts in Chinese literature as far back as 3,300 B.C.E.

Shun-Sheng uses these historical antecedents, along with practical measurements of the use of the centerboard on a bamboo raft, to argue that the idea for the centerboard originated in China and spread across the North Pacific ultimately to the balsa wood rafts of the western coast of South America. The 4,000-year-old jar showing an ancient Mesopotamian centerboard found by Thor Heyerdahl notwithstanding, these are apparently the only two places in the world where centerboard navigation was employed before 1870. Unless one assumes independent invention, the idea originated either in China and spread to America along prevailing North Pacific currents or, as Heyerdahl argued, it originated amongst the balsa raft sailors of South America, to be carried by prevailing winds and currents to Asia (see Heyerdahl's *American Indians in the Pacific*, 1952, 593).

As for the scientific substance of de Bisschop's reverse *Kon-Tiki* expedition, Ben Finney would correctly appraise it years later as "valiant, if anthropologically misguided" (Finney, et al., 1994, 272).

Edwin Ferdon visited Henderson Island during Thor Heyerdahl's East Pacific archaeological expedition in 1957, and found a single human skeleton in repose there (see Ferdon, *One Man's Log*, Chicago, Rand McNally, 1966, 137). Several more human skeletons were noted in 1958. These were removed by an 1991-92 expedition, and analyzed by Dr. Marshall Weisler, who identified them as conclusively prehistoric, with further details pending. As far as the seemingly strange building material used to construct *Tahiti-Nui II*, it is entirely likely that had rafters landed in South America from Polynesia in prehistoric times, they would have been forced to employ local materials for a canoe or raft for the return voyage.

William Willis: Age and Adventure

Willis during his first leg makes some interesting speculations about the prehistoric navigators who preceded him across the Pacific:

"The night was clear, and sitting by the compass, I thought of the ancient Polynesians making their way over these same seas in their big double canoes. How their navigators must have gazed up at the constellations, measuring as best they could the height of the large stars above the horizon, aware that if they calculated wrong they might never see land again. There was, besides, the ever-present danger of squalls and headwinds, of gales and shifting currents and days and nights of cloudy weather, to wipe out all possibility of seeing the sun or stars and so computing their position. Once lost, there was no turning back, since they didn't know which way to turn, not having a compass [here Willis ignores the star compass above his head]; they had to sail on, even when food and water had given out and the dying had begun" (Willis 1966, 104).

Willis' 1954 balsa raft *Seven Little Sisters* seems to have died an even more gruesome death. Perhaps the government of American Samoa thought the old raftsman was never coming back to their islands, or perhaps rabid Samoanists, if such people exist, resented the implication that their islands could have been settled by white-haired drifters from Peru. For whatever reason, when Willis returned to Pago Pago from his forced landing in independent Western Samoa in November of 1963, he found that the raft was nowhere to be found. He had been warned in Western Samoa that the people in the eastern islands had chopped it up for firewood, but could not bring himself to believe it. After all, Willis had donated the raft to the U.S. government and that government had accepted it in perpetuity. If any other arrangements were ever made for it, the agreement required that Willis or his heirs be notified. Willis himself, like the philosopher at heart he was, seemed to accept the raft's fate:

> …I walked over to the Government House and stood on the lawn where the *Seven Little Sisters* had been. It was empty. Sunk without trace—but not sunk flying the Stars and Stripes till the waves closed over her brave body, for she had been a fighter and had sailed to a world's record through seven thousand unbroken miles of storms and calms in 115 grueling days. That's how she would have wanted to die… (ibid, 170).

Age Unlimited was brought to the Mariners' Museum in Newport News, Virginia, where it awkwardly stuck out amongst the typological collections of world maritime history. The museum in the 1970s went through a policy of determining the significance of the each item in its vast collection: what could be kept, restored, and permanently preserved, and what was inappropriate and/or beyond restoration. According to William D. Wilkinson (pers. comm., 26 July 2000), director of the museum from 1972-94, the criterion used for keeping an artifact in the small boat collection was its importance to the development of well-known small craft types. Willis' raft, deteriorating to the point where it was hardly recognizable, was both outside the scope of the collection and prohibitively expensive to restore. Even had restoration been affordable, Willis himself was by now deceased, lost at sea during an attempted transatlantic expedition, and no drawings or sketches remained upon which the curators could have based a restoration. The raft was finally dismantled and discarded.

The boat upon which Willis' lost his life, Little One, was later retrieved by a Soviet trawler, handed over to the U.S. Navy, and also brought to the Mariners' Museum. It too was eventually deemed beyond restoration and discarded. According to William B. Cogar (e-mail, 3 August 2000), Vice President and Chief Curator of the Mariners', selected artifacts from both vessels were kept by the museum, including "binoculars, port holes, camera, articles of clothing used and worn by Willis, plus a painting of Little One."

Heyerdahl takes on the Atlantic Ocean

Heyerdahl's experiments with *Ra* and *Ra II* are chronicled in *The Ra Expeditions* (Garden City, New York: Doubleday, 1971), a book that may even eclipse *Kon-Tiki* as the most brilliant account of a maritime expedition ever written. It is equally gripping to those who love their anthropological canvasses painted in broad, global strokes. Heyerdahl himself comes across as the ultimate expedition companion—generous, humorous, good-natured, always ready to take a tough turn. The only kind of man who could make one answer "yes" to a voyage across an ocean on a boat made from paper. I have never met anyone who has read this book who has not come away from it without wishing they could have been on board for the ride. Indeed one, Phil Buck, built his own reed boat and sailed it to Easter Island in the year 2000 because he had read *The Ra Expeditions* as an eleven-year-old boy.

It was my great pleasure to spend a day in Woods Hole, Massachusetts, with the navigator of Ra, *Ra II*, and *Tigris*, Captain Norman Baker. Sitting on the edge of Eel Pond, we had the opportunity to cover dozens of aspects of these reed boat voyages, which have taken their rightful place as some of the most important archaeological experiments of the twentieth century. Norman Baker is serious, open, and so obviously rock-steady that after five minutes with him it becomes obvious why Heyerdahl placed so much responsibility upon his shoulders. He told the unforgettable story of the vote taken on board *Ra II* during our talk in Woods Hole. He told it as well during the *Kon-Tiki* + 50 dinner held at the Cosmos Club by the Philadelphia and Washington Chapters of the Explorers Club in April 1997. It is recounted in an article written after that dinner: "Target Thor Heyerdahl," (The Explorers Journal 75 (1), Summer 1997). During our talk in Woods Hole, I had a chance to question Baker more closely about the character of Thor Heyerdahl. He described him as "the most straightforward, open, friendly guy you'd ever want to meet. And he's that way to this day. It's one thing for a wealthy man to give away money; but to be generous with your praise, with your compliments, with your own personal time, that is an entirely different form of generosity. We all of us have only so many days, and for him to give his time to someone, that is generosity."

Heyerdahl's Last Great Experiment

Mohenjo-Daro, the "Mound of the Dead," is Pakistan's most famous archaeological site. The site is one of more than 1,000 located within the Indus Valley, center of the Harrappan civilization dating from 2,500 B.C.E.—1,700 B.C.E. The Indus Valley drains the Western Himalayas, an area larger than that of Egypt and Mesopotamia combined, and was intensively explored in 1921-22 by Sir John Marshall and the Archaeological Survey of India. The sites of Harrappa and Mohenjo-Daro are both located in Pakistan. Lothal, the presumed port city of Harrappan civilization, is located over the border in India.

A mural at the national museum in Bahrain, which I saw in 2003, depicts a reed ship flotilla arriving at the main gate of the city thousands of years ago, presumably from connecting these great civilizations. The voyage of the *Tigris* is covered in Heyerdahl's book *The Tigris Expedition: In Search of Our Beginnings* (New York: Doubleday, 1980). The performance of the raft exceeded all predictions, and at one point Norman Baker even managed to force it 80° into the wind.

Closing the Argument: Chile to Easter Island

Phil Buck describes the sailing characteristics of the voyage of the *Viracocha* as follows:

"The total straight-line distance between Arica and the Easter Island is just over 2,400 nautical miles. Our course was nearly direct and I estimated we sailed about 2,500 miles to get there. For about two weeks (weeks three and four) in a row we were able to average ninety miles a day, with a few days over one hundred. The main course and wind changes occurred in the final two weeks (weeks five and six). After being becalmed near Sala Y Gomez island, the wind picked up again and blew directly from astern and helped us sail right into Easter Island, again on fifteen foot seas. It was quite a ride! Overall we were able to sail at 75 degrees against the wind at our most efficient with both sails flying and 35 degrees to it under the main only, which we only discovered as we neared Easter Island. I believe that our voyage was the only reed boat voyage that truly was not a drift voyage" (Phil Buck, personal communication, August 14, 2000).

The Way to Central America

Data on the *Illa Tiki*, *La Manteña-Huancavillca I* and *II* , *La Enduriencia* expeditions was provided through several personal talks with John Haslett over many years, several pieces of expedition literature provided by Haslett to the author as well as the expedition's web-site: http:??www.balsaraft.com. The expedition was featured on the ABCNews web-site, and in an Associated Press article datelined in Bogota, Columbia, which appeared in the Philadelphia Inquirer of Wednesday, January 6, 1999, under the headline, "Raft sets off again on trip across time," by Jared Kotler. The tragic finale of *La Enduriencia* is seen in a two-part television documentary, produced for the Adventure Quest series by American Adventure Productions, Inc., which appeared under the title, "*La Manteña: Rafting Across the Pacific*." Detailed descriptions of the construction and sailing characteristics of the raft are contained in archaeologist Cameron McPherson Smith's excellent preliminary report on the voyage, *Manteño Expedition, 1998-1999* (London: Royal Geographical Society, May, 1999).

The fascinating reference to possible Norse knowledge of teredo-like marine invertebrates during the time of Eirik the Red, who colonized Greenland in the year 985 C.E., is from Eirik's Saga, part thirteen. See The Vinland Sagas: The Norse Discovery of America, translated and with an introduction by Magnus Magnusson and Hermann Pálsson. (New York: Penguin Books, 1965, 103.)

The Norwegians return to the Pacific

This chapter is based largely upon personal interviews I conducted with Torgeir Higraff in July of 2004 and again in the summer of 2007.

The other experiments after *Kon-Tiki*: Science

Savoy's account of the *Kuviqu* expedition can be found in *On the Trail of the Feathered Serpent* (Indianapolis: The Bobbs—Merrill Company, Inc., 1974).

Despite the adoption of similar reed boat technology to what Thor Heyerdahl was then using for his summer, 1969, voyage in *Ra I*, Savoy barely mentions Heyerdahl in his account. When he does, he misstates at least one of Heyerdahl's positions on prehistoric voyaging. At one point, discussing the origin of the Olmecs, Savoy writes that: "Heyerdahl suggested that the Olmecs were Nordic adventurers from across the sea" (Savoy 1974, 10). Since this absurd "suggestion" is not attributed, it is impossible to know where Savoy found it. But it is interesting in that it is typical of many off-hand, casual, and ridiculous thoughts wrongly attributed to Thor Heyerdahl. Savoy then launches into a discussion of the many different races apparently present in the New World at the time of the Europeans. He finds no "Old World" solution to this apparent dilemma—and hence no need to look for transoceanic contacts—obtusely deciding that: "Americans were simply here, possibly before the continents began to drift apart" (ibid.). A sea drift of a very different kind indeed!

Even more than DeVere Baker, it is difficult to know just where Savoy's religion ends and his archaeology begins, or vice versa. He made astounding discoveries in the 'antisuyu' region of the Inca empire (the jungle areas east of the Andes), in the 1960s and the 1980s, by following his own daring and ultimately correct hypothesis that both Incan and pre-Incan societies had occupied areas of the Amazon basin rain forest. His allusions to Jesus in the retelling of the Viracocha legend, however, draw one to the section of his website that describes the explorer as "The Most Right Reverend Douglas Eugene Savoy, Head Bishop of the International Community of Christ, Church of the Second Advent." His biography adds to this impressive title those of "President of the Jamilian University of the Ordained; Chancellor of the Sacred College of Jamilian Theology; President of the Advocates for Religious Rights & Freedoms; [and] Chairman of the World Council for Human Spiritual Rights." Savoy is listed as well as the founder of a trademarked religion called Cosolargy "a system of healing for the spirit and soul."

After the *Kuviqu*, or *Feathered Serpent*, voyage, Savoy organized *Feathered Serpent II*, a schooner expedition from 1977 to 1982 to study winds and currents that may have carried prehistoric mariners between Asia and America. He followed this with an expedition in a 73-foot mahogany catamaran, *Feathered Serpent III*, ("a replica of designs found on pre-Inca ceramics and of traditional Polynesian craft" (Savoy website)) from Callao to Hawai'i from December 17, 1997 to January 28, 1998. This latter voyage was organized when Savoy believed he had found archaeological evidence linking cave sites in the Andes with the biblical city of Ophir, toward which King Solomon directed ships in search of gold for his Jerusalem temple. Like Baker before him, he decided that this proposition could best be explored by a cruise to Hawai'i.

Like Heyerdahl a remarkably resilient explorer, Savoy continues to make headlines with his archaeological discoveries in the almost-impenetrable jungles of the eastern slopes of the Andes.

See, for example, "Cities in the Sand," U.S. News & World Report, July 10, 2000.

Primary sources for the *La Balsa* and Las Balsas expeditions can be found in Vital Alsar, *La Balsa* (New York: Reader's Digest Press, 1973); and Vital Alsar, ¿Por que impossible? Las Balsas (Barcelona: Editorial Pomaire, 1976). For access to a copy of the latter publication, I am grateful to Maja Bauge, Director of the Kon-Tiki Museum in Oslo.

Information on the *Uru* expedition was found on the website of the publication *Aula de el Mundo* (January 20, 2000), in an article entitled "Expedicion *Uru*," and translated from the Spanish using Alta Vista's Babel Fish translation software. Data on the launch of the *Mata Rangi I* comes from an article in the *Rapa Nui Journal* 11(1), 43); from a CNN web article dated December 31, 1996, entitled "Easter Island sails into the past," by CNN Santiago Bureau Chief Ronnie Lovler; and from an article about a visit by the King of Spain to the *Mata Rangi* in 1997, in *Te Rapa Nui*, the Gazette of Easter Island 11 (3), 1. Paulino Esteban, who oversaw the construction of the massive *Mata Rangi I* on the shores of Easter Island's [pre] historic Anakena Bay, also had some interesting comments on the origin of the Easter Island reeds, pointing out that pollen samples collected from the crater lakes—ironically by Heyerdahl's 1955 expedition—seem to show that the species was introduced to an uninhabited Easter Island by birds some 30,000 years ago. If correct, it would seem to preclude their introduction by reed boat mariners sailing from the coast of Peru a thousand years ago. See "Sailing with Paulino," (*Te Rapa Nui* 11 (3), 4).

I am indebted to Ben Finney for passing along the latter article, as well as newspaper notes about the *Mata Rangi II.*

Perhaps because it was seen as a failure, little has been written about the voyage of the *Chimok*, although a brief television documentary was made of the expedition. (I am grateful to John Haslett for providing me with a copy.) Certainly it was fortuitous that the German crew decided to conduct the experiment at the same moment when Thor Heyerdahl was excavating at Túcume, and could benefit from his experience and advice. Heyerdahl mentions the *Chimok* in Heyerdahl, et al., *Pyramids of Túcume: The Quest for Peru's Forgotten City* (New York, Thames & Hudson, 1995, 221-223).

Material describing the *Titi* expedition is in "Around the Lake in 28 Days," (Bolivian Times, Vol. 11, No. 46, 9, 18-25 November 1994), and "Around Lake Titicaca on a Reed Boat in 28 Days," (Bolivian Times, Vol. 1, No. 10, 3, 8 October 1993). I found these references during research at the Kon-Tiki Museum in Oslo in August 1999.

Information on the *Mata Rangi* expeditions comes from an exhibit at the Kon-Tiki Museum in Oslo. Additional details appeared as well in articles that appeared on the web at the time of the voyage, including "Spaniard bids to cross Pacific Ocean in reed boat," a Reuters article dated February 14, 1999, and datelined in Arica, Chile.

I learned of the *Abora* expedition (and its predecessor expeditions *Dilmun I, Dilmun II,* and *Dilmun III*), during research at the Kon-Tiki Museum in Oslo in August 1999.

Besides the long-term storage of the reeds, which likely affected their buoyancy, the Leipzig reeds were very slim. One person familiar with the project told me that Heyerdahl himself did not have confidence in this species of reed when used as the basis of a sailing vessel. The mast was also stepped too far aft, negatively affecting the reed boat's sailing characteristics.

Colonel Blashford-Snell recently e-mailed me and noted that the reed flagship from the *Kota-Mama* II expedition is now housed at the ISCA Maritime Museum in Lowestoft, England, where reside a large collection of rafts and reed boats. "She is still drying out, having weighed some eight tons at the start [and] 21.5 at the end after some seventy days afloat" (John Blashford-Snell, personal communication, 7 Aug. 2000).

The other experiments after *Kon-Tiki*: Pseudo-Science

Thor Heyerdahl cited many of these same Olmec sculptures, decidedly without the outrageously racist overtones, to argue for possible transatlantic contacts prior to the voyages of the reed boats *Ra I* and *Ra II* in 1969 and 1970. (See Heyerdahl, *Early Man and the Ocean*, New York: Doubleday & Co., 1979, 121). Heyerdahl also brought up "Uncle Sam" as he discussed isolationism versus diffusionism with Norman Baker as the *Ra I* drifted across the Atlantic:

"[One type of Olmec representation] has a well-defined, sharp profile, with a strong aquiline nose, small, thin-lipped mouth and often a moustache and goatee or flowing full beard. The archaeologists have jokingly nicknamed this type 'Uncle Sam.' 'Uncle Sam' is generally portrayed with a majestic head-dress, full-length cloak, belt and sandals. This type, strongly semitic in appearance and often carrying a wanderer's staff, is illustrated from the Olmec area southward as far as the legends about the white men go. Modern religious sects have often cited this in support of their belief in the 'lost tribes of Israel' or the holy 'Book of Mormon'" (Heyerdahl 1971, 251).

DeVere Baker claimed that the Tres Zapotes ceramic head dated to about the time of Christ, but in fact it is likely from a period much earlier than that. In the 1950s, radiocarbon tests at La Venta produced a range of dates from 900-400 B.C.E. A new series of dating at San Lorenzo during the 1980s produced an even earlier range of dates spanning 1150-900 B.C.E.

Santiago Genovés wrote of his unique transatlantic drift expedition in *The Acali Experiment: Five Men and Six Women on a Raft Across the Atlantic for 101 Days* (New York, Times Books, 1980). The sociological research is written up in a separate section of the book that runs from page 297-444.

Tim Severin and *Hsu Fu*

Tim Severin is second only to Thor Heyerdahl himself in the brilliance of his writing. He has a well-deserved international reputation for authoring intensely interesting books about fascinating subjects. For anyone interested in recreated maritime technology and transoceanic contacts, a single chapter of Severin often contains more nautical details than one generally finds in whole ethnographic monographs. I was profoundly affected by Severin's book *The Brendan Voyage* (reprint edition: New York: Modern Library, 2000), about a crossing of the North Atlantic in a leather boat, which I read as an undergraduate student more than twenty years ago. It was one of the books that convinced me that a life in exploration would be the most interesting life possible. It seems that there is no project of historic (or prehistoric) interest, which Severin has not thought of, if not actually carried out. As a doctoral student, I remember asking one of my professors, one with considerable research interests in the Southwest Pacific, whether or not anyone had recreated the famous canoe voyage of Alfred Russell Wallace. Wallace had recounted the expedition in his natural history classic The Malay Archipelago, and I thought that a modern recreation of that voyage would be a fascinating project. The professor didn't know, but also thought it was a good idea. It couldn't have been more than a few months later when I heard that Tim Severin was already on the case. He produced another immense classic, *The Spice Islands Voyage* (New York: Carroll & Graf, 1997). The route has also been admirably retraced by Australian science writer Penny Van Oosterzee in *Where Worlds Collide: The Wallace Line* (Ithaca: Cornell Univ. Press, 1997).

The China Voyage (paperback edition: Reading, Massachusetts: Addison-Wesley Publishing, 1995) continued a remarkable series of voyages on recreated technology and books about these expeditions. Furthermore, Severin was explicit in conducting what he himself called an "archaeological experiment" (Severin 1995, 156). Whereas Heyerdahl and others had proven that a solid balsa logs raft had the ability to drift "downhill" with the winds, no one had attempted an intercontinental voyage over the much more treacherous North Pacific in a flexible bamboo raft.

By a seeming coincidence, Severin asked the Mariners' Museum in Newport News, Virginia—the same museum where William Willis had deposited *Age Unlimited*—if they would accept his bamboo raft for display should it reach America. The museum did not have the space to accession yet another transoceanic raft, but it did arrange for a grant to link the raft with local school classrooms that followed *Hsu Fu*'s progress across the Pacific. The abandoning of the raft, of course, rendered any museum display discussions moot (ibid, 71).

The Origins of the Raft

The Joseph Birdsell excerpt comes from his article "The Recalibration of a Paradigm for the First Peopling of Greater Australia," in J. Allen, et al., Sunda and Sahul: *Prehistoric Studies in Southeast Asia, Melanesia and Australia*, London: Academic Press, 1977).

The recent discoveries and very recent dating of the *Homo erectus* skulls to between 27,000 and 53,000 years ago were reported in an article by Swisher III, et al, in Science 274 (5294), 1870-1874. Many of the questions related to the varying geographic distribution of and human population and arrival in Sahul, including Birdsell's article, are included in J. Allen, et al., Sunda and Sahul: *Prehistoric Studies in Southeast Asia, Melanesia and Australia*, London: Academic Press, 1977).

There is evidence that Indian traders knew of Java as early as 600 B.C.E., and Hindu and later Buddhist influence was strong from about 200 C.E. until the 1400s, when Islam began displacing Hinduism. The Hindu influence was felt predominantly in the upper classes, while the rural classes retained their traditional animistic cults. Hindu legal practices were brought to Indonesia, as were the levels of social rank and ceremony that prevail even today. The Hindi concept of a divine ruler with unlimited powers is especially strong in Indonesia. Indonesians made pilgrimages as far as Tibet to absorb more philosophy and metaphysics. Buddhism became dominant on Sumatra, while Hinduism predominated on Java.

The last great Java-Hindu state, which is also considered the Golden Age of Indonesia, lasted from 1294 to 1398, and its ruler, Gajah Mada, was said to have worked so tirelessly that it took four officials to do his job when he died. During this period, native folk art was revitalized and broke through the old Hindu-Buddhist models. Even today, four hundred years after Islam displaced Hinduism in the islands, the culture and religion of Bali is inherited directly from India.

Arab traders had arrived in the islands of Indonesia as early as the fourth century after Christ. They followed a route from Africa and India that led through Southeast Asia and China, bringing oil palm and the kapok tree from Africa. Soon after Islam began its spread through parts of India, it began its journey to Indonesia, displacing all previous religions, first in Sumatra, then on Java.

Islam taught the common man, who had existed previously with feudal lords wielding absolute powers, that in the eyes of Allah all men were created from the same clay (women, of course, being a separate story entirely). According to one account, Islam is the perfect religion for such an island nation, combining the virtues of sound commercial law, prosperity, and hard work, while simultaneously acting as a political bulwark against Portuguese and later Dutch domination.

Muslim merchants were scattered by the Portuguese capture of Melaka in 1511. The Portuguese, however, were not as interested in conquering territory as they were in controlling commercial markets. When the Portuguese murdered the Sultan of the island of Ternate in 1570, hoping for a better deal with his successor, they were themselves thrown out by the natives, and this began the decline of their influence in the islands.

Portuguese rule was then challenged and rapidly eclipsed by the ascendant Dutch, who landed four ships in the islands in 1596. When these four ships returned safely to Holland with cargo holds full of spices, Dutch trading concerns went wild, sending a dozen expeditions with sixty-five ships to the Indies during the seven years from 1598-1605.

It is difficult to overstate the amounts of wealth accumulated by the Dutch East India Company, or its worldwide influence on the course of human history. Formed in 1602, the company rapidly obtained an exclusive monopoly on Dutch trade east of the Cape of Good Hope and west of the Strait of Magellan. By law, no one could sail into the East Indies from either Africa (by way of the Cape of Good Hope), or South America (by way of Cape Horn), without the permission of the Dutch East India Company. This edict was occasionally defied, however, and the Dutch sea captain Schouten sailed into the Pacific in 1611 and reached Batavia, after sailing past what are today the Schouten Islands and Biak off the northern coast of New Guinea.

The company claimed and exercised the power to make treaties, build forts, garrison troops, and impose courts of law throughout the islands. The company's ruthless agents eventually drove the British and Portuguese from Indonesia and dominated trade with the Spice Islands (Moluccas). When corruption and mismanagement forced its collapse in 1798, its markets and possessions became part of the Dutch Empire in East Asia.

The Dutch then replaced the corporation with a massive bureaucracy of civil servants, replacing a commercial apparatus with a colonial one. An enforced cultivation system was put in place in 1830, solely to feed the demand in Europe for coffee, sugar, tea, pepper, and cotton. According to one account, virtually the entire island of Java was turned into one enormous forced labor camp, similar to slave plantations in the American South before the Civil War. The Dutch established an ironclad colonial system that regarded the Indonesians as sub-humans. Swimming pool signs read "No Natives or Dogs," and no higher education was made available until the 1920s. At the time the flying boat that we searched for in 1997 sank at Manokwari—just prior to the Japanese invasion at the start of the Second World War—ninety per cent of the population was still illiterate.

The Origins of the First Rafters

This episode occurs in one of Heyerdahl's lesser-known classics, *The Maldive Mystery* (Bethesda, Maryland: Adler & Adler, 1986, 187-190). Photos of the bamboo raft, and the Burmese candy wrapper, appear in a gather of photos between pages 88-89. It would make for an interesting ethnological fieldtrip, indeed, to trace the course of this bamboo vessel back to Burma, and recount the lives of the raft builders there. Is there any local tradition of using such large cargo-carrying rafts on deepwater voyages over several thousand miles employing, as Heyerdahl speculated, the winds of the Northeast Monsoon to propel the raft across the waters of the Bay of Bengal? If so, it would make for an interesting journey to explore the ancient routes of Buddhist mariners voyaging out from the coast of Burma. If one went by statuary alone, Buddha himself could be Heyerdahl's original Long-Ear.

Heyerdahl's speculation more than half a century ago that evidence exists for massive human migrations in prehistory is increasingly borne out by modern scientific work. His career in the Pacific has been dominated by the idea that balsa raft drifters were forced to sea sometime around 500 C.E., by either a cultural or a climatic disaster, or perhaps the former triggered by the latter. Heyerdahl's study of the cradle of civilization in the Middle East and his experimental reconstructions of the *Ra I*, *Ra II*, and *Tigris*, center on the notion of a catastrophic disturbance of the ancient world that he sees occurring around 3,000 B.C.E.

Evidence for a massive climatic event around 530 C.E., one that destabilized cultures along the north Peruvian coast (and perhaps around the globe), is found both in tree ring and in ice core data. The archaeological journalist David Keys, in a brilliant synthesis, argues that the likely cause of this rapid and devastating climate shift was an explosion of the Krakatoa volcano in what is now the Sunda Straits separating the Indonesian islands of Java and Sumatra. Keys argument is contained in Catastrophe: An Investigation into the Origins of the Modern World (New York: Ballantine Books, 2000).

Others have argued that the evidence of the demise of the Moche points toward a severe El Niño Southern Oscillation (ENSO) event. Heyerdahl himself raised this possibility in a letter to me several years ago. He had witnessed first-hand the effects of the severe El Niño flooding of the Peruvian coast in 1983, when balsa rafts could be seen drifting on flood waters several miles from their coastal ports. An excellent synthesis of the evidence for the destabilizing effects of El Niño is contained in Brian Fagan's book, Floods, Famines, and Emperors: El Niño and the Fate of Civilizations (New York: Basic Books, 1999). Evidence that a flood of truly biblical proportions decimated the ancient world approximately 5,000 B.C.E., comes from recent research by Walter Pitman and William B. F. Ryan, and chronicled in Noah's Flood: The New Scientific Discoveries About the Event That Changed History (New York: Touchstone, 1999).

Finally, as this book goes to press, comes word (September 2000) of new and dramatic discoveries by University of Pennsylvania archaeologist Fred Heibert and undersea explorer Robert Ballard of sites of human occupation submerged more than 300 feet below the surface of the Black Sea. Such a level seems to indicate a massive flood event in the area some 7,000 years ago, one that could have led to the biblical legends of a great flood.

ACKNOWLEDGEMENTS

I have visited the Kon-Tiki Museum four times since 1993, and was assisted there by the director, Maja Bauge, during a research visit in August 1999. Former Director, Dr. Arne Skjølsvold, answered many questions I put to him via letter in the spring of 2000, and I spent most of a day with Norman Baker in July 2000. Withal, I made no effort to contact Dr. Heyerdahl directly. Having utilized his voluminous writings, visits to his museum, correspondence with Dr. Skjølsvold and personal interviews and e-mails with Norman Baker, I felt it unnecessary to interrupt the legend himself as he made the most of his latter years of research.

Thor Heyerdahl passed away in 2002 and his influence on young explorers has seemingly grown since that time. Both the *Tangaroa* expedition in 2006 and the 2012 cinema film *Kon-Tiki* have generated new enthusiasm for the study of the many archaeological puzzles generated by Thor Heyerdahl's lifetime of experimental transoceanic raft expeditions.

In Kon-Tiki's Wake is the result of a lifetime of interest in rafts and rafting, and three years of intensive reading about such experiments and incorporating this material into my "Introduction to Archaeology" course at Penn State University's Abington College. Since no general history of transoceanic raft expeditions exists, I entrusted myself with the task of producing one. At Abington College, I am grateful to former Associate Dean Leonard Mustazza and former Head of the Division of Social and Behavioral Sciences James F. Smith for their support and especially for making possible my participation at the 1999 "Symposium on Maritime Achaeology and History of Hawaii and the Pacific" in Honolulu. The Librarians at Abington, especially Jeanette Ullrich, tracked down dozens of obscure references over the course of two years. Michael Aaron Rockland, former chair of the American Studies Department of Rutgers University, introduced my work to Leslie Mitchner at Rutgers University Press, who in turn thought the transoceanic raft history idea worth pursuing, and for that I will always be grateful. The project has now taken on a new life as a paperback in Fireship Press' series of works in nautical non-fiction with the help of Michael James.

I have drawn inspiration along the way from many expeditions and many people. None were so impressive as Dr. Ben R. Finney, whom I had the honor of meeting at the Maritime Archaeology and History Conference in Honolulu, Hawaii, in 1999, and Captain Norman Baker, whom I was likewise privileged to interview in Woods Hole, Massachusetts, during the summer of 2000. Both have shared stories, articles, and books with me, all of which made my job easier as they enriched the content of my prose. As with all who provided information and anecdotes, it goes without saying that all strengths of a book like this are theirs, all errors of content or interpretation solely my own.

The publisher Ed Burlingame offered me the irresistible opportunity to write a foreword to The Adventure Library's 50th anniversary edition of *Kon-Tiki* in 1997, and produced what I believe is the most beautiful edition of Thor Heyerdahl's classic of exploration. That subsequent raft and double hull canoe voyages have already rendered my introduction obsolete by no means diminishes the wonderful production of the TAL edition.

David Herdrich, of the American Samoa Historic Preservation Office in Pago Pago, very kindly tracked down several references to the landing of William Willis' rafts there in 1954 and 1963. This included a picture of the *Seven Little Sisters* as it was displayed on the island after Willis donated it to the people of American Samoa. William Wilkenson, former director, and William B. Cogar, current curator, of the Mariners' Museum in Newport News, Virginia, helped me sort out the fate of Willis' *Age Unlimited.*

Through letters, e-mails, data from websites, telephone conversations and personal meetings, several current raft explorers shared their experiences. These included Colonel John Blashford-Snell, Phil Buck, John Haslett, and Torgeir Higraff. Phil Buck, who has lived the expedition I so long dreamed of—a reed boat voyage from northern Chile to Easter Island—was both generous and kind in letting me use material from his exemplary *Viracocha* expeditions.

I must also say a special word about my friends, the raft explorers John Haslett and Torgeir Higraff. I first met John as he prepared for the 1997 voyage of *La Manteña-Huancavillca.* Since then, we have shared periodic and intense conversations of all topics related to prehistoric raft voyages. Torgeir's generosity with his experiences, and his almost superhuman endurance, make him, like John Haslett and also Phil Buck, yet another dynamic explorer following so ably in the wake of Thor Heyerdahl's *Kon-Tiki.*

FURTHER STUDY NOTES FOR IN KON-TIKI'S WAKE: THE ARCHAEOLOGY OF THE TRANSOCEANIC EXPERIMENT

Part 1: Introduction to experimental archaeology

The archaeologist is a scientific explorer. His laboratory is the planet earth, the moon, Mars, anywhere that humans have left evidence of their past behavior. Archaeology is the extraction of scientific data from the examination of artifacts and other evidence of human behavior. All fields of archaeology—whether the study of ancient Egypt or the study of discarded spacecraft on the moon —share a basic approach to observation and analysis. They all derive data from artifacts and use this data to create arguments or hypotheses concerning human behavior.

The experimental archaeologist is a scientific adventurer. Experimental archaeologists evaluate their hypotheses through the creation and use of an artifact thought to have been in use in the past. Often, this involves recreating a maritime vessel from the past and putting to sea on a literal and figurative voyage into the past with one's own life at risk in the process.

After the Second World War, the Norwegian anthropologist Thor Heyerdahl was challenged to demonstrate his notion that a raft made out of balsa wood could drift from South America to the islands of eastern Polynesia. Taking up this challenge, Heyerdahl conceived of the *Kon-Tiki* expedition as a scientific test of an anthropological hypothesis. Nothing of the sort had ever been done before. A seagoing archaeological experiment.

First, Heyerdahl had to recreate a vessel he believed prehistoric peoples had invented and mastered nearly two millennia into the past. Second, he then proposed to risk his life by making an actual voyage on board this vessel, which he believed could remain afloat for months. Finally, Heyerdahl proposed to use his recreated prehistoric raft to connect a continent with an archipelago separated by more than 4,000 miles of ocean.

Heyerdahl's balsa raft, which did not sink to the bottom of the Pacific as so many had predicted it would, is now the centerpiece of perhaps the most unique museum in the world: the only institute anywhere that is devoted in the main to a single archaeological experiment.

Experimental maritime archaeology after *Kon-Tiki*

Since the *Kon-Tiki* expedition in 1947, dozens of other maritime archaeological experiments have been undertaken. You will now begin to read and study these expeditions, and witness several films about them. These include a Mesopotamian reed ship sailed through the Persian Gulf, and Polynesian double-hulled voyaging canoes sailed along legendary routes from Hawai'i to Tahiti and across much of the rest of Polynesia.

These latter experiments, conducted by University of Hawai'i anthropologist Ben R. Finney, have shed light on several aspects of prehistoric voyaging with Polynesian double-hulled canoes. For example, a longstanding notion held that "prevailing" winds are a permanent barrier to windward navigation. Finney has shown that a properly-sailed Polynesian deepwater canoe can either sail successfully against the Pacific trade winds, or sit tight to await seasonal shifts in the prevailing winds in order to reach their destinations.

By the late 1980s, at a time when Heyerdahl was nearing eighty, a new generation of scientists was eager to venture into the field with him. In the years before he passed away in 2002 at the age of eighty-seven, Heyerdahl kept himself as relevant to questions of intense archaeological interest as when he was thirty-six.

By then, a new generation of raft expeditions had begun to set out in search of answers to finely tuned anthropological questions. How did the balsa raft manage to voyage so far across the Pacific without the balsa being attacked and eaten by *teredo navalis* worms? Was it the time of harvest that made them more or less resilient against this destructive wood-borer? Or did floating the logs down the freshwater rivers from the interior balsa stands of Ecuador provide some protection from the wood borers until the raft was safely at sea and perhaps beyond their reach? How did prehistoric cultures deal with the teredo?

One thinks of Tim Severin noting that the raft sailors of Vietnam beach their rafts each night on a full tide, leaving their hulls high and dry and beyond the reach of the teredo. Or John Haslett, speculating that perhaps prehistoric mariners used tar pits or seeps along the Pacific coast to apply a protective layer of tar to their logs. These intellectual descendents of Heyerdahl are not just extending experimental archaeology, but forging whole new fields, like paleomarine-ethnobiology, to track the evolution and attempt to time the appearance of marine invertebrates and their affects on human explorations of the Pacific.

To make their efforts even more difficult, these explorers must conduct their experiments in the absence of the novelty of *Kon-Tiki*. Like the astronauts who followed Apollo 11, they might gain notoriety for a moment, but never again will a raft expedition be considered unique. Heyerdahl took all of the risks and gained all of the rewards of the pioneer. It was six years before another explorer set out on another raft in imitation of *Kon-Tiki*. By contrast, in the six years between 1993 and 1999, seven different raft expeditions were organized and carried out. Those raft experiments also operated in a far different popular culture environment, one of saturation news, extreme sports, and a minute collective attention span.

"Seeing if it's possible..."

These archaeological experiments owe much to the example of the *Kon-Tiki* expedition. The sailor Eric de Bisschop once wrote that Thor Heyerdahl was the only academic he respected. He said this was because, unlike most anthropologists who moved prehistoric human populations around on theoretical chessboards, Heyerdahl was alone in taking his ideas onto real oceans, learning true lessons.

The navigator on several of Heyerdahl's expedition, Captain Norman Baker, has remarked that Heyerdahl was never about demonstrating whether or not some population went from here to there. "What Thor [was] doing [was] testing, seeing if it's possible," said Baker. "The fact that we crossed the Atlantic in *Ra* doesn't prove that it was done in ancient times, it only means that it was possible."

This latter thought is the cornerstone of Heyerdahl's rightful place as the founder of maritime experimental archaeology. This subfield of the study of the past takes problems out of the library and places them in real world 'laboratory' settings. Where Heyerdahl's true greatness lies is in the fact that he transformed all of the planet's seas and oceans into archaeological laboratories.

Archaeologists can argue about how one can place strict scientific controls on the experiments emanating from these laboratories. But the fact remains that such archaeological experiments are not testing the chemical structure of a molecule. They are attempting to understand human behavior, which in itself is infinite and variable.

Even in such a vastly different world than the one that greeted *Kon-Tiki* with such enthusiasm in 1947, each of the new generation of raft explorers takes Heyerdahl as their inspiration. And it is more than simply his experimental model. His generous and optimistic nature drew generation after generation of explorers to seek him out.

Norman Baker, who voyaged with Heyerdahl on three reed boats across more than 10,000 miles of ocean, remarked that he would set sail on another raft tomorrow if there was something new to be learned. "Thor changed my vision of the world. His expeditions defy the imagination. We arrived in Barbados on board *Ra II*, after two months at sea, with twelve hours of buoyancy left. If you wrote a screenplay about a man who set out to cross an ocean on a boat made from paper it would be dismissed as fantasy."

Key terms

Berdi totora
balsa
teredo navalis
kumara/sweet potato
ethnoarchaeology
experimental archaeology
moai No. 263
Wallace Line

Key personalities

Thor Heyerdahl
Ben Finney
John Haslett
David Lewis
Andrew Sharp
Phil BuckEric de Bisschop
Arne Skjølsvold
Tim Severin
John Blashford-Snell
Maximo Catari
Alain Bombard
Torgeir Higraff
John Haslett

Part 2: Sample exam questions

Match the expedition leader to his sea route

1. Ben Finney	a. Chile to Easter Island
2. David Lewis	b. China to America
3. Phil Buck	c. Hawai'i to Tahiti
4. Eric de Bisschop	d. Tahiti to New Zealand
5. Tim Severin	e. Tahiti towards South America

Match the experimental maritime vessel to the type of material that was used in its construction:

1. *Titi*	a. berdi reed
2. *Kon-Tiki*	b. totora reed
3. *Tigris*	c. bamboo
4. *Hsu-Fu*	d. papyrus reed
5. *Ra II*	e. balsa logs

Match the experimental maritime vessel to the experimental sea route or place most closely associated with them from your readings and classroom lectures:

1. *Hokule'a*	a. Chile to Easter Island
2. *Ra II*	b. China to America
3. *Viracocha*	c. Hawai'i to Tahiti
4. *Tahiti-Nui*	d. Africa to the Americas
5. *Hsu-Fu*	e. Tahiti to Chile

Short essay questions

1. What was the effect on Heyerdahl's thinking of his year-long stay in the Marquesas?
2. Why has kumara excited the interest of Pacific scholars?
3. How might the Humboldt Current be changed to bring mariners to Easter Island?
4. What are the challenges to an archaeologist trying to link prehistoric societies via maritime routes.

5. What was it about the *Kon-Tiki* experiment that so vexed Eric de Bisschop?

6. How did Bisschop envision the geographic dispersions of the Polynesians?

7. What are some of the lessons that can be taken from the drift of *Tahiti-Nui II*?

8. What was the dilemma caused by Moai 263 and how did Heyerdahl attempt to resolve it experimentally?

9. *Ra II*, built by Aymara Indians, was twenty feet shorter than *Ra I* and presumably closer in construction to a prehistoric reed ship than *Ra I.* Still, the raft began to take on water as soon as it entered the sea, sinking how much per week in the first two weeks of the expedition?

10. With the decks of *Ra II* awash, Heyerdahl still faced a journey of how many miles across the Atlantic, in his attempt to link the Old World with the New World?

11. How does the guara seem to link Asia and America?

12. What did Severin discover about the nature of a transoceanic bamboo raft?

13. What does the *Chimok* experiment say about all archaeological raft experiments?

14. How did Haslett attempt to defeat the teredo after the destruction of *Illa-Tiki*?

15. The experiment of Maximo Catari bears on two distinct yet related archaeological questions. One is the possible transport of large stonework via reed ships. What is the other?

16. Why was passing close by Sala-y-Gómez so important to Phil Buck during the *Viracocha* expedition?

17. Does the success of *Tangaroa* and the way it was achieved change the view of experimental raft expeditions? If so, how?

Sample Questions for *Kon-Tiki: Across the Pacific by Raft* by Thor Heyerdahl

Chapter 1

1. Heyerdahl describes the trade winds as "eternal." How else might they be viewed?

2. Heyerdahl had made "the identification of the legendary hero Tiki" the objective of his archaeological and anthropological research. Why?

3. Heyerdahl writes that "the original Polynesian race must at some time, willingly or unwillingly, have come drifting or sailing to these remote islands." Which scenario did Heyerdahl himself believe in?

Chapter 2

1. Heyerdahl's friend Wilhelm told him that: "Perhaps ten [prehistoric] rafts went to the bottom for every one that got across —or perhaps hundreds in the course of centuries." What implications does this notion have for archaeological studies?

2. In theoretically ideal conditions, with a fair wind, and assuming the raft would sail as expected, Heyerdahl's friend Wilhelm calculated the *Kon-Tiki* expedition across the Pacific would take about ninety-seven days. How long did he say they should be prepared to remain at sea if his calculations were off?

Chapter 3

1. Why was Don Federico so encouraging when asked about the capabilities of the balsa raft?

2. As if to reinforce the connection he is trying to make between South America and Polynesia, Heyerdahl christened each of his twelve balsa logs with what names?

3. Heyerdahl writes that *Kon-Tiki* is a faithful copy of an early Peruvian or Ecuadoran balsa raft with the exception of which feature, which he declares was found to be completely unnecessary anyway?

Chapter 4

1. The authenticity or experimental aspect of the *Kon-Tiki* expedition can be questioned on a number of levels. What non-prehistoric element is present right at the start of the expedition?

2. *Kon-Tiki* was "built exactly as the Spaniards described it," writes Heyerdahl. As an experimental archaeologist, why was this only half of his problem?

Chapter 5

1. What possible sea routes to Polynesia does Heyerdahl offer for the legendary figure of *Kon-Tiki*, following his postulated defeat at Lake Titicaca in prehistory?

2. According to Heyerdahl, the original prehistoric rafters who left Peru were supplied with dried meat and fish, sweet potatoes, and water stored primarily in the skins of giant bottle gourds and ________?

Chapter 6

1. Midway in his journey, Heyerdahl remarks that he no longer believes in the sea as a what?

2. Heyerdahl speculates that a group of "long-ears" settled Easter Island, or Rapa Nui, around the years 400-500 AD. Why does he believe this might be important in the overall settlement of Polynesia?

Chapter 7

1. The most immediate danger of landing the twenty-ton raft at Raroia Island in the Tuamotu Archipelago was what?

2. The second danger of landing at Raroia, if you survived the first, was what?

Study questions for *The Tigris Expedition*, a *National Geographic* film by Dale Bell

In 1977-78, explorer Thor Heyerdahl led one last great archaeological experiment, on board the reed boat *Tigris*. This pivotal experiment collated and exploited much of the ethnographic, ethnohistorical, and archaeological data on the construction of prehistoric reed boats that Heyerdahl had uncovered over the course of forty years of international research and experimentation.

The following are some questions for you to consider as you watch the *National Geographic* television special made from the expedition by film producer Dale Bell. You must also read Chapter 18 in Sea Drift, the chapter that encapsulates the history of the expedition.

1. Prior to the *Tigris* expedition, Thor Heyerdahl learned that, in order to be fully buoyant, the berdi reed growing in the marshes of Iraq must be harvested when?

2. During the *Tigris* expedition, Heyerdahl learned that a Mesopotamian reed ship like *Tigris* could theoretically carry a cargo estimated at how many tons?

3. Heyerdahl was forced to bring South American Aymara indians to Iraq to build the *Tigris* because of what complication?

4. By locating and studying cuneiform tablets and Indus Valley script on the island of Bahrain, archaeologist Geoffrey Bibby, interviewed by Thor Heyerdahl during the *Tigris* expedition, seemingly found evidence for prehistoric contacts between ________________ and ____________.

5. The three civilizations of the Old World that Heyerdahl linked through the *Tigris* expedition were those of the ______________, ________________, and ________________.

Study questions for "The Voyage of the Hokule'a," a *National Geographic* Film by Dale Bell

A second documentary film, also made by Dale Bell, recorded the voyage of the Hawai'ian replica canoe *Hokule'a*. This project was led by Ben Finney of the University of Hawai'i along with colleagues such as Herb Kawainui Kane of the Polynesian Voyaging Society (http://pvs.kcc.hawaii.edu).

1. What kind of vessel did Ben Finney design and build and how did he use it to explore prehistoric Polynesian navigation?

2. What were the unintended consequences of the experiments?

3. Why did the construction of the canoe produce so much tension between the designers of the experiment and the workers building the canoe?

4. Were the tensions of the construction phase eliminated once the canoe was launched? Why or why not?

APPENDIX I

A chronology of experimental and other raft voyages

Year	Name	Type	Route	Leader
1947	*Kon-Tiki*	Balsa Raft	Peru to Raroia	Thor Heyadahl
1952	*L'Heretique*	Rubber Raft	Europe to Caribbean	Alain Bombard
1954	*Seven Sisters*	Balsa raft	Peru to American Samoa	William Willis
1954	*Lehi*	Plywood raft	San Francisco toward Hawai'i	DeVere Baker
1955	*Lehi II*	Plywood raft	San Francisco toward Hawai'i	DeVere Baker
1955	*La Cantuta*	Balsa raft	N. Peru to Galapagos	Eduardo Ingris
1956-7	*Tahiti Nui*	Bamboo raft	Tahiti to Chile	Eric De Bisschop
1957	*Lehi III*	Plywood raft	San Francisco to Los Angeles	DeVere Baker
1958	*Tahiti Nui II*	Cedar raft	Chile to near Starbuck Island	Eric de Bisschop
1958	*Tahiti Nui III*	Metal drum	Starbuck Island to Rakahanga	Eric de Bisschop
1958	*Lehi IV*	Plywood raft	Redondo Beach, Cal. to Hawai'i	Eric de Bisschop
1958	*L'Egare II*	Cedar raft	Newfoundland to Ireland	Henri Beaudout
1959	*La Cantuta II*	Balsa raft	Peru to Matahiva	Eduardo Ingris

IN KON-TIKI'S WAKE

Year	Name	Type	Route	Leader
1963	*Age Unlimited*	Balsa raft	Peru to Samoa and Australia	William Willis
1965	*Tangaroa*	Balsa raft	Central Peru to Fakareva	Carlos Caravedo Arca
1969	*Feathered Serpent*	Reed boat	Peru to Panama	Gene Savoy
1969	*Celeusta*	Rubber raft	Central Peru to Raroia	Mario Valli
1969	*Ra*	Reed boat	Africa toward America	Thor Heyerdahl
1970	*Ra II*	Reed boat	Balsa raft	Ecuador to Australia
1973	*La Aztlan*		Ecuador to Australia	Vital Alsar
	La Guayaquil			
	La Mooloolaba			
1973	*Acali*	Wooden raft	Europe to Caribbean	S. Genovese
1974	*Tai Ki*	Bamboo raft	China Sea toward Ecuador	Kuno Knöbl
1978	*Tigris*	Reed boat	Mesopotamia to Africa	Thor Heyerdahl
1988	*Uru*	Reed boat	Peru to Kitin	Muñoz Marquesas
1993	*Hsu Fu*	Bamboo raft	Taiwan toward America	Tim Severin
1995	*lla Tiki*	Balsa raft	Ecuador to Panama	John Haslett
1997	*Mata Rangi*	Reed boat	Easter Island toward Asia	Kitin Muñoz

Year	Name	Type	Route	Leader
1998	*La Mantena*	Balsa raft	Ecuador to Costa Rica	John Haslett
1998	*Kota-Mama I*	Reed boat	Desaguadero River	John Blashford-Snell
1999	*La Enduriencia*	Balsa raft	Costa Rica	John Haslett
1999	*Mata Rangi II*	Reed boat	Chile toward Asia	Kitin Muñoz
2000	*Viracocha*	Reed boat	Chile to Easter Island	Phil Buck
2006	*Tangaroa*	Balsa raft	Peru to Polynesia	Torgeir Higraff

APPENDIX II

Experimental raft expeditions by motivation

Science	Pseudo-science	Adventure	Survival
Kon-Tiki	*Lehi I*	*Seven Little Sisters*	*L'Heritique*
Tahiti-Nui	*Lehi II*	La Cantuta *I*	*Tahiti-Nui III*
Feathered Serpent	*Lehi III*	*L'Egare I*	
Ra	*Lehi IV*	*L'Egare II*	
Ra II	*Tahiti-Nui II*	*La Cantuta II*	
La Balsa	*Acali*	*Age Unlimited*	
La Tres Balsas		*Pacifica*	
Tigris		*Celeusta*	
Uru		*La Enduriencia*	
Hsu Fu			
Illa-Tiki			
Chimok			
Mata-Rangi			
La Manteña			
Abora			
Titi			
Kota-Mama			
Viracocha			
Tangaroa			
Total: 20	6	9	2

APPENDIX III

The Expeditions of Archaeological Significance

Rafts	Method	Significance
Kon-Tiki	Balsa	Establishes experimental nautical archaeology as a sub-field; supports hypothesis that prehistoric maritime vessels built from balsa wood have global reach.
Tahiti-Nui	Bamboo	A bamboo raft can survive for up to six months, even in the mid-latitudes of the South Pacific Ocean.
Ra II	Papyrus	A crescent-shaped reed ship of prehistoric design can cross an ocean.
La Balsa	Balsa	A balsa raft can drift across the entire Pacific Ocean from South America to Australia.
Tigris	Berdi reed	A reed ship of ancient design can remain afloat for six months or more, and link the ancient culture centers of Mesopotamia, Dilmun, Indus Valley and Egypt. Demonstrates the efficacy of the circumferential rope in reed ship construction.
Hsu Fu	Bamboo	The Chinese junk-style bamboo raft is capable of crossing an ocean. It is not the bamboo that wears down, but the rattan lashings that hold the raft together that must be replaced often.
Mata Rangi	Totora reed	A three-masted reed vessel twice the length of *Ra II* and based directly on the carving of *Moai* 263, it worked itself to pieces very quickly and sank less than 200 miles from its launch point on Rapa Nui.

Illa-Tiki	Balsa	A balsa raft used in coastal waters is open to rapid destruction by *Teredo navalis* worms.
La Manteña	Balsa	A balsa raft attempting to cross from South to Central America risks capture and destruction in the Gyre west of Panama.
Viracocha	Totora reed	A reed ship can make a rapid passage connecting the culture area of the Inca/pre-Inca with Rapa Nui (Easter Island). Confirms that the circumferential rope is essential in the construction of the reed ship.
Tangaroa	Balsa	A guara-equipped balsa raft held on a straight-line course in the Humboldt Current can cross from Peru to Polynesia in just two months, an entire month faster than *Kon-Tiki* in 1947.

APPENDIX IV

Where are the rafts now?

Raft Location

<u>1940s</u>

1. *Kon-Tiki* On display, Kon-Tiki Museum, Oslo, Norway

<u>1950s</u>

1. *L'Hérétique* Unknown

2. *Seven Little Sisters* Made into a museum in American Samoa in 1954; on display for an indeterminate length of time. Raft was gone by the fall of 1963.

3. *La Cantuta* Unknown

4. *Lehi I* Lost at sea.

5. *Lehi II* Lost at sea.

6. *Lehi III* Lost at sea.

7. *Lehi IV* Abandoned in Hawai,i.

8. *L'Egare I* Broken up and sunk, Nova Scotia, 1953?

9. *L'Egare II* Unknown

10. *Tahiti Nui* Broken up and sunk off the coast of Chile, 1956

11. *Tahiti Nui II* Broken up and sunk north of Flint Island, 1957

12. *Tahiti Nui III* Broken up and sunk on Rakahanga Reef, 1957

13. *La Cantuta II* Unknown

1960s

1. *Age Unlimited* Donated to Mariner's Museum, Norfolk, Virginia; stored unmaintained out-of-doors until being discarded in the 1970s.

2. *Tangaroa* Unknown

3. *Pacifica* Unknown

4. *Celeusta* Unknown

5. *Feathered Serpent* Destroyed by storm in Bay of Panama, 1969

6. *Ra I* Sunk 500 miles east of Barbados, 1969

1970s

1. *Ra II* On display, Kon-Tiki Museum, Oslo, Norway

2. *La Balsa* Reported on display at maritime museum in Santander, Spain, 1970

3. *La Aztlan* On display at Maritime Museum, Bellina, Australia

4. *La Guayaquil* Unknown

5. *La Mooloolaba* Unknown

6. *Acali* Unknown

7. *Tai Ki* Broken up and sunk in the North Pacific, 1974

8. *Tigris* Burned in protest of warfare in the Middle East, off the coast of Somalia, 1978

1980s

1. Burmese bamboo raft Broken up and auctioned in Viringili, Maldives

2. *Uru* Unknown

1990s

1. *Hsu Fu* Broken up and sunk in the North Pacific, 1993
2. *Illa Tiki* Abandoned on shore in Panama, 1995
3. *Chimok* Abandoned off the coast of Ecuador
4. *Mata Rangi* Broken up and abandoned 300 miles from Easter Island
5. *Mata Rangi II* Unknown
6. *La Manteña* Broken up and abandoned off the coast of Costa Rica
7. *La Enduriencia* Broken up on the rocks off coast of Costa Rica, 1999
8. *Abora* Unknown
9. *Titi* On display at Catari family museum in Huatajata, Bolivia
10. *Kota-Mama* On display at the ISCA Maritime Museum, Lowestoft, England

2000s

1. *Viracocha* Burned at Easter Island
2. *Tangaroa* In active use at Larvik, Norway

BIBLIOGRAPHY

Allen, J., et al. 1977. Sunda and Sahul: *Prehistoric Studies in Southeast Asia, Melanesia and Australia*. London: Academic Press.

Alsar, Vital. 1973. *La Balsa*. New York: Reader's Digest Press.

________. 1976. ¿Por que impossible? Las Balsas. Barcelona: Editorial Pomaire.

Bahn, Paul., and John Flenley. 1992. *Easter Island Earth Island*. New York: Thames and Hudson.

Baker, DeVere. 1959. *The Raft Lehi IV*. Long Beach, California: Whitehorn Publishing.

Baker, Nola. 1960. *The Raft Dog*. Salt Lake City: Bookcraft.

Baker, Devere, and Nola Baker. n.d. *The Raft, The Meteorite, and a Dog!* Privately printed.

Baker, Norman, with Barbara Beasley Murphy. 1974. *Thor Heyerdahl and the Reed Boat Ra*. Philadelphia: J.B. Lippincott.

Baker, Norman. 1997. "Target Thor Heyerdahl." *The Explorers Journal* 75 (1), summer.

Beaudout, Henri. 1957. *The Lost One*. London: Hodder & Stoughton, 1957.

Bednarik, Robert G. 2003. "Seafaring in the Pleistocene." *Cambridge Archaeological Journal* 13 (1): 41-66.

Birdsell, Joseph. 1977. "The Recalibration of a Paradigm for the First Peopling of Greater Australia." In J. Allen, et al. Sunda and Sahul: *Prehistoric Studies in Southeast Asia, Melanesia and Australia*. London: Academic Press.

Birkett, Dea. 1997. *Serpent in Paradise*. New York: Doubleday.

Bisschop, Eric de. 1959. *Tahiti Nui*. New York: McDowell, Oblensky.

Bombard, Alain. 1953. *The Bombard Story*. London: Andre Deutsch.

Bower, Bruce. 2003. "*Erectus* Ahoy: Prehistoric Seafaring Floats into View," *Science News* 164 (18): 248–250.

British Broadcasting Corporation. 1999. 'First Americans were Australian.' http://news.bbc.co.uk/2/hi/science/nature/430944.stm (accessed January, 2013).

Buck, P.H. 1938. *Vikings of the Sunrise*. New York: F.A. Stockes.

Capelotti, P.J. 1997. *Introduction to Kon-Tiki, by Thor Heyerdahl*. North Salem, New York: The Adventure Library.

________. 1997b. "The Elusive Island." *The Explorers Journal* 75 (1), summer.

________. 2007. *Flåteferder i Kon-Tikis kjølvann*. Oslo: Flyt forlag.

________. 2012. "The Theory of the Archaeological Raft: Motivation, Method, and Madness in Experimental Archaeology," EXARC: *The International Journal of Archaeological Open Air Museums and Experimental Archaeology*: 8/2.

"Cities in the Sand." 2000. U.S. News & World Report, July 10.

Claassen, Cheryl. 1983. "Answering Our Questions with Experiments," in Richard A. Gould, ed. *Shipwreck Anthropology*. Albuquerque: University of New Mexico Press.

Conrad, Joseph. 1975 (1896). *The Outcast of the Islands*. New York: Penguin Books.

Cowan, James. 1996. *A Mapmaker's Dream: The Meditations of Fra Mauro, Cartographer to the Court of Venice*. Boston: Shambhala.

Cox, Paul Alan., and Sandra Anne Banack, eds. 1991. *Islands, Plants, and Polynesians: An Introduction to Polynesian Ethnobotany*. Portland, Oregon: Dioscorides Press.

Danielsson, Bengt. 1960. *From Raft to Raft*. London: George Allen & Unwin.

Darwin, Charles. 1989 (1839). *Voyage of the Beagle*. (Edited by Janet Browne and Michael Neve.) New York: Penguin Books.

Dewan, Leslie and Dorothy Hosler. 2008. "Ancient Maritime Trade on Balsa Rafts: An engineering analysis." *Journal of Anthropological Research* 64 (1): 19-40.

Dodge, Ernest. 1969. "Review of Sea Routes to Polynesia by Thor Heyerdahl," *Man*, new series 4 (1): 162-63.

Ferdon, Edwin N. 1966. *One Man's Log*. Chicago: Rand McNally.

________. 1968. "Polynesian Origins." in Andrew Vayda, ed. *People and Cultures of the Pacific*. New York: Natural History Press.

Finney, Ben R., ed. 1976. *Pacific Navigation and Voyaging*. Wellington, New Zealand: The Polynesian Society Incorporated.

________. 1979. *Hokule'a: The Way to Tahiti*. New York: Dodd, Mead & Co.

________. 1992. *From Sea to Space*. Palmerston North, New Zealand: Massey University.

________. 1993. Lecture. University of Pennsylvania Museum, November 12, 1993.

Finney, Ben R., Paul Frost, Richard Rhodes, and Nainoa Thompson. 1989. "Wait for the West Wind." *Journal of the Polynesian Society* 98.

Finney, Ben R., et al. 1994. *Voyage of Rediscovery*. Berkeley: University of California Press.

Genovés, Santiago. 1980. *The Acali Experiment: Five Men and Six Women on a Raft Across the Atlantic for 101 Days*. New York: Times Books.

Gladwin, Thomas. 1970. *East is a Big Bird: Navigation & Logic on Puluwat Atoll*. Cambridge, Massachusetts: Harvard University Press.

Golson, J. 1968. "Review of "Reports of the Norwegian Archaeological Expedition to Easter Island and the East Pacific," *Man*, new series 3 (2): 322-323.

Harrison, H.S. 1953. "Review of Thor Heyerdahl's American Indians in the Pacific: The Theory Behind the Kon-Tiki Expedition," *Man* 53: 45-48.

Haslett, John. 2006. *Voyage of the Manteno: The Education of a Modern-Day Expeditioner*. New York: St. Martin's Press.

Heyerdahl, Thor. 1941. "Did Polynesian Culture Originate in America?" New York: *International Science* 1.

________. 1950. *Kon-Tiki*. Chicago: Rand McNally & Company.

________. 1952. *American Indians in the Pacific: The Theory behind the Kon-Tiki Expedition*. London: George Allen and Unwin.

________. 1958. *Aku-Aku: The Secret of Easter Island*. London: George Allen and Unwin.

________. 1971. *The Ra Expeditions*. Garden City, New York: Doubleday.

________. 1974. *Fatu Hiva: Back to Nature*. Garden City, New York: Doubleday.

________. 1979. *Early Man and the Oceans*. Garden City, New York: Doubleday.

________. 1980. *The Tigris Expedition*. Garden City, New York: Doubleday.

________. 1986. *The Maldive Mystery*. Bethesda, Maryland: Adler & Adler.

________. 1996. *Green Was the Earth on the Seventh Day*. New York: Random House.

Heyerdahl, Thor, and Edwin N. Ferdon, Jr., eds. 1961. *The Archaeology of Easter Island: Reports of the Norwegian Archaeological Expedition to Easter Island and the East Pacific*, Volume 1. Santa Fe, New Mexico: Monographs of the School of American Research and the Museum of New Mexico, number 24, part 1.

________. 1966. *The Archaeology of Easter Island: Reports of the Norwegian Archaeological Expedition to Easter Island and the East Pacific*, Volume 2. Santa Fe, New Mexico: Monographs of the School of American Research and the Museum of New Mexico, number 24, part 2.

Heyerdahl, Thor, Daniel H. Sandweiss, and Alfredo Narvaez. 1995. *Pyramids of Túcume: The Quest for Peru's Forgotten City*. New York: Thames and Hudson.

Higraff, Torgeir. 2006. "Tangaroa expedition newsletters." http://tangaroa.nettblogg.no/english.html (accessed January, 2013).

Hornell, J. 1946. "How did the Sweet Potato reach Oceania?" Journal of the Linnean Society , Vol. LIII, No. 348.

Johansen, Øystein Kock, ed. 1999. *Norwegian Maritime Explorers and Expeditions Over the Past Thousand Years*. Oslo: Index Publishing.

Jones, Rhys. 1977. "Sunda and Sahul: An Introduction," in J. Allen, et al. Sunda and Sahul: *Prehistoric Studies in Southeast Asia, Melanesia and Australia*. London: Academic Press.

Kane, Herb Kawainui. 1976. "A Canoe Helps Hawai'i Recapture Her Past." *National Geographic* 149 (4), April.

Kirch, Patrick Vinton. 1996. *The Lapita Peoples: Ancestors of the Oceanic World*. Oxford: Blackwell.

________. 2000. *On the Road of the Winds*. Berkeley: University of California Press.

Kraus, Bob. 1988. *Keneti: South Seas Adventures of Kenneth Emory*. Honolulu: University of Hawai'i Press.

Lewis, David. 1972. *We, the Navigators: The Ancient Art of Landfinding in the Pacific*. Honolulu: University Press of Hawaii.

________. 1976. "Hokule'a Follows the Stars to Tahiti." *National Geographic* 150 (4), October.

Melville, Herman. 1992 (1852). *Moby-Dick*. Introduction by Andrew Delbanco and Notes by Tom Quirk. New York: Penguin Books.

Metraux, Alfred. 1957. *Easter Island: A Stone-Age Civilization of the Pacific*. London: Andre Deutsch.

Michener, James A. 1982 (1959). *Hawaii*. Paperback edition. New York: Fawcett Crest.

Moore, Thomas. 1990. "Thor Heyerdahl: Sailing Against the Current," U.S. News & World Report, April 2.

Murray, John A., ed. 1991. *The Island and the Sea: Five Centuries of Nature Writing from the Caribbean*. New York: Oxford University Press.

Oosterzee, Penny Van. 1997. *Where Worlds Collide: The Wallace Line*. Ithaca: Cornell University Press.

Rensch, Karl H. 1991. "Polynesian Plant Names: Linguistic Analysis and Ethnobotany, Expectations and Limitations," in Paul Alan Cox and Sandra Anne Banack, eds. *Islands, Plants, and Polynesians: An Introduction to Polynesian Ethnobotany*. Portland, Oregon: Dioscorides Press.

Ryan, Donald. 1997. "Thor Heyerdahl: Explorer, Scholar, and World Citizen." *The Explorers Journal* 75 (1), summer.

Savoy, Gene. 1974. *On the Trail of the Feathered Serpent*. Indianapolis: The Bobbs-Merrill Company, Inc.

Severin, Tim. 1995. *The China Voyage*. Paperback edition. Reading, Massachusetts: Addison-Wesley.

________. 1997. *The Spice Islands Voyage*. New York: Carroll & Graf.

________. 2000 (1978). *The Brendan Voyage*. Reprint edition. New York: Modern Library.

Sharp, Andrew. 1963. *Ancient Voyagers in Polynesia*. Berkeley: University of California Press.

Shun-Sheng, Ling. 1956. "[The] Formosan Sea-going Raft and its Origin in Ancient China." Bulletin of the Institute of Ethnology Academia Sinica 1 (March): 2-54.

Smith, Cameron McPherson. 1999. *Manteño Expedition, 1998-1999.* London: Royal Geographical Society.

Swift, Jonathan. 1981 (1726). *Gulliver's Travels.* Introduction by Miriam Kosh Starkman. New York: Bantam Books.

Swisher III, C.C., et al. 1996. "Latest *Homo erectus* of Java: Potential Contemporaneity with Homo sapiens in Southeast Asia." *Science* 274 (5294): 1870-1874, December 13.

Thiel, Barbara. 1987. "Early Settlement of the Philippines, Eastern Indonesia, and Australia-New Guinea: A New Hypothesis." *Current Anthropology* 28 (2): 236-251.

Thorpe, Nick. 2002. *8 Men and a Duck: An Improbable Voyage by Reed Boat to Easter Island.* New York: The Free Press.

Trigger, Bruce G. 1989. *A History of Archaeological Thought.* Cambridge: Cambridge University Press.

Vayda, Andrew P. 1968. "Polynesian Cultural Distributions in New Perspective." In Andrew P. Vayda, ed. *People and Cultures of the Pacific.* New York: Natural History Press.

Wandless, Edgar G. 1943. *The Story of the Rubber Life Raft.* New York: New York Rubber Corporation.

Ward, R. Gerard, John W. Webb, and M. Levison. 1976. "The Settlement of the Polynesian Outliers: A Computer Simulation." In Ben R. Finney, ed. *Pacific Navigation and Voyaging.* Wellington, New Zealand: The Polynesian Society Incorporated.

Whistler, W. Arthur. 1991. "Polynesian Plant Introductions." In Paul Alan Cox and Sandra Anne Banack, eds. *Islands, Plants, and Polynesians: An Introduction to Polynesian Ethnobotany.* Portland, Oregon: Dioscorides Press.

Willis, William. 1955. *The Gods Were Kind.* New York: E.P. Dutton.

William Willis. 1966. *Whom the Sea Has Taken.* New York: Meredith.

________. 1969. *The Hundred Lives of an Ancient Mariner.* London: Hutchinson.

About the Author
P.J Capelotti

P.J. Capelotti earned his B.A. (1983) and M.A. (Phi Kappa Phi, 1989) degrees in history from the University of Rhode Island, and an M.A. (1994) and Ph.D. (1996) in anthropology from Rutgers University. He wrote the introduction to the 50th anniversary edition of *Kon-Tiki* published by The Adventure Library in 1997, and is the author of more than a dozen books, including *Shipwreck at Cape Flora: The Expeditions of Benjamin Leigh Smith, England's Forgotten Arctic Explorer* (Calgary, 2013). He teaches Archaeology and American Studies at Penn State University Abington College in Abington, Pennsylvania.

For the Finest in Nautical and Historical Fiction and Nonfiction

Interesting • Informative • Authoritative

CPSIA information can be obtained at www.ICGtesting.com
Printed in the USA
BVOW05s0239240715

410252BV00008BA/87/P

9 781611 792690